Biological and Medical Physics, Biomedical Engineering

More information about this series at http://www.springer.com/series/3740

BIOLOGICAL AND MEDICAL PHYSICS, BIOMEDICAL ENGINEERING

The fields of biological and medical physics and biomedical engineering are broad, multidisciplinary and dynamic. They lie at the crossroads of frontier research in physics, biology, chemistry, and medicine. The Biological and Medical Physics, Biomedical Engineering Series is intended to be comprehensive, covering a broad range of topics important to the study of the physical, chemical and biological sciences. Its goal is to provide scientists and engineers with textbooks, monographs, and reference works to address the growing need for information.

Books in the series emphasize established and emergent areas of science including molecular, membrane, and mathematical biophysics; photosynthetic energy harvesting and conversion; information processing; physical principles of genetics; sensory communications; automata networks, neural networks, and cellular automata. Equally important will be coverage of applied aspects of biological and medical physics and biomedical engineering such as molecular electronic components and devices, biosensors, medicine, imaging, physical principles of renewable energy production, advanced prostheses, and environmental control and engineering.

Chang-Hwan Im
Editor

Computational EEG Analysis

Methods and Applications

Springer

Editor
Chang-Hwan Im
Department of Biomedical Engineering
Hanyang University
Seongdong-gu, Seoul, South Korea

ISSN 1618-7210 ISSN 2197-5647 (electronic)
Biological and Medical Physics, Biomedical Engineering
ISBN 978-981-13-0907-6 ISBN 978-981-13-0908-3 (eBook)
https://doi.org/10.1007/978-981-13-0908-3

Library of Congress Control Number: 2018946584

This Springer imprint is published by the registered company Springer Nature Singapore Pte Ltd.
The registered company address is: 152 Beach Road, #21-01/04 Gateway East, Singapore 189721, Singapore

Preface

The electroencephalogram (EEG), a recording of electrical activities in the brain, is becoming an indispensable tool to investigate human brain functions and to diagnose various psychiatric and neurological disorders. Since the first recording of a human EEG by Dr. Hans Berger, a psychiatrist, in 1924, the development of EEG technology has continued, and, increasingly, this technology has drawn interest from researchers in various disciplines, including clinicians, neuroscientists, psychologists, and biomedical engineers. Owing to recent advances in digital technology and software methodology, EEG is now being used as an important tool in numerous fields, such as cognitive neuroscience, neuromarketing, neuroergonomics, brain–computer interfaces, neurofeedback, and sports science.

Although there is a consensus that EEGs are easier to record than other brain-imaging techniques, such as functional magnetic resonance imaging and positron emission tomography, the analysis of EEGs is not straightforward. For example, if one wants to observe functional connectivity between two brain regions of interest, it is necessary to perform a series of EEG processing steps, including pre-processing, EEG source imaging, and functional connectivity analysis. Although there are several software packages offering comprehensive tools for advanced EEG analyses, users still need to choose the specific computational EEG analysis methods most appropriate for their EEG data. Indeed, there are many kinds of methods for computational EEG analysis, e.g., a variety of functional connectivity measures. Therefore, it is recommended that EEG researchers understand the detailed theoretical background of the computational EEG analysis methods being used. Knowledge of the advantages and disadvantages of each method would help researchers achieve more successful EEG analysis results.

In this book, we intend to provide a comprehensive review of the state-of-the-art methods for computational EEG analysis. This book is not a handbook, but a textbook written by multiple experts. Therefore, this book should be useful not only to biomedical engineers who are in the initial stages of working on the development of EEG analysis methods, but also to clinicians/neuroscientists who want to acquire extensive knowledge of the technical details of the EEG analysis tools that they use for their research. We hope the readers find this book useful.

Seoul, South Korea Chang-Hwan Im

Contents

Contributors

Jeong Woo Choi Department of Biomedical Engineering, Yonsei University, Wonju, South Korea

Marco Congedo GIPSA-Lab, Centre National de la Recherche Scientifique (CNRS), Grenoble-INP, Université Grenoble Alpes, Grenoble, France

Chang-Hwan Im Department of Biomedical Engineering, Hanyang University, Seoul, South Korea

Garett D. Johnson Department of Electrical and Computer Engineering, Old Dominion University, Norfolk, VA, USA

Ki-Young Jung Department of Neurology, Seoul National University, Seoul, Republic of Korea

Do-Won Kim Department of Biomedical Engineering, Chonnam National University, Yeosu, South Korea

Kyung Hwan Kim Department of Biomedical Engineering, Yonsei University, Wonju, South Korea

Sung-Phil Kim School of Design and Human Engineering, Ulsan National Institute of Science and Technology, Ulsan, Republic of Korea

Dean J. Krusienski Department of Electrical and Computer Engineering, Old Dominion University, Norfolk, VA, USA

Seung-Hwan Lee Department of Psychiatry, Inje University, Ilsan-Paik Hospital, Goyang, Republic of Korea; Clinical Emotion and Cognition Research Laboratory, Inje University, Goyang, South Korea

Yeonsoo Park Clinical Emotion and Cognition Research Laboratory, Inje University, Goyang, South Korea

Dan Zhang Department of Psychology, School of Social Sciences, Tsinghua University, Beijing, China

Yingchun Zhang Department of Biomedical Engineering, University of Houston, Houston, USA

Abbreviations

ADC	Analog-to-digital converter
AR	Autoregressive or autoregression
BCI	Brain–computer interface
BEM	Boundary element method
BSS	Blind source separation
CAR	Common average reference
CCA	Canonical correlation analysis
CDS	Cortically distributed source
CSD	Current source density
CSF	Cerebrospinal fluid
CSP	Common spatial pattern
DCM	Dynamic causal modeling
DFT	Discrete Fourier transform
dSPM	Dynamic statistical parametric mapping
DTF	Directed transfer function
DWT	Discrete wavelet transform
ECD	Equivalent current dipole
EEG	Electroencephalogram or electroencephalography
EMD	Empirical model decomposition
EMG	Electromyogram
EOG	Electrooculogram
ERD	Event-related desynchronization
ERP	Event-related potential
ErrP	Error-related potential
ERS	Event-related synchronization
ESI	Electrical source imaging
FEM	Finite element method
FFT	Fast Fourier transform
FIR	Finite impulse response
fMRI	Functional magnetic resonance imaging

fNIRS	Functional near-infrared spectroscopy
GC	Granger's causality
GLM	General linear model
HFO	High-frequency oscillation
ICA	Independent component analysis
IED	Interictal epileptiform discharge
iEEG	Intracranial EEG
IIR	Infinite impulse response
IMF	Intrinsic mode function
ISI	Inter-stimulus interval
LAURA	Local autoregressive average
LMS	Least mean squares
LORETA	Low-resolution electromagnetic tomography
LTI	Linear time-invariant
MAP	Maximum a posteriori
MEG	Magnetoencephalography
MI	Motor imagery (in brain–computer interfaces)
MI	Mutual information (in functional connectivity analysis)
MMN	Mismatch negativity
MNE	Minimum-norm estimates
MVAR	Multivariate autoregression model
NVC	Neurovascular coupling
OCD	Obsessive compulsive disorder
PCA	Principal component analysis
PDC	Partial directed coherence
PET	Positron emission tomography
PLI	Phase lag index
PLV	Phase locking value
PTSD	Post-traumatic stress disorder
RLS	Recursive least squares
sLORETA	Standardized low-resolution electromagnetic tomography
SNR	Signal-to-noise ratio
SOZ	Seizure onset zone
SPECT	Single photon emission tomography
SQUID	Superconducting quantum interference device
STFT	Short-time Fourier transform
TE	Transfer entropy
wMNE	Weighted minimum-norm estimates

Part I
Introduction

Chapter 1
Basics of EEG: Generation, Acquisition, and Applications of EEG

Chang-Hwan Im

Abstract The purpose of this chapter is to provide comprehensive knowledge about the generation and acquisition of electroencephalograms (EEGs), which is essential for understanding the following chapters. The physiological background on the generation of EEGs is presented, and then, a detailed description of the acquisition of EEG signals is given. Practical applications of computational EEG analysis are also introduced. Finally, the major advantages and limitations of current EEG technologies are discussed.

1.1 Generation of EEG

An electroencephalogram (EEG) is the flow of neuronal ionic currents recorded using a pair of electrodes either inside or outside the scalp. The EEG signal recorded inside the skull, referred to as the intracranial EEG (iEEG), can be used for surgical planning of intractable epilepsies [15]; however, this is not dealt with in this book (except in Chap. 8). Throughout this book, "EEG" refers to a *scalp EEG* recorded noninvasively from a pair of electrodes attached to the scalp surface.

In comparison with brain metabolism- or hemodynamics-based neuroimaging modalities, such as positron emission tomography (PET), functional magnetic resonance imaging (fMRI), and functional near-infrared spectroscopy (fNIRS), EEGs can offer excellent temporal resolution, allowing studies of neuronal dynamics occurring within a few milliseconds. However, the spatial resolution of an EEG is not comparable to that of an fMRI, owing to the small numbers of spatial data samplings, inherent volume conduction effect, and physiological and environmental noises/artifacts.

A first human EEG was recorded in 1924 by a German psychiatrist, Hans Berger. Despite the rapid technological developments, the basic methods for recording EEGs remain unchanged from Hans Berger's era. An EEG measures electric potential differences between pairs of electrodes. The electrodes may be either directly attached

C.-H. Im (✉)
Department of Biomedical Engineering, Hanyang University, Seoul, South Korea
e-mail: ich@hanyang.ac.kr

C.-H. Im (ed.), *Computational EEG Analysis*, Biological and Medical Physics, Biomedical Engineering, https://doi.org/10.1007/978-981-13-0908-3_1

to the scalp surface at some specific locations or fitted in a cap (or a net) for more convenient attachment.

The main generators of the EEG, often referred to as *EEG sources*, are cortical neurons. It is well-documented that most neurons in the human brain are concentrated within the cerebral cortex, which is a thin sheet of gray matter with 2–4 mm thickness. The apical dendrites of the cortical neurons, often referred to as *large cortical pyramidal neurons*, are arranged almost perpendicularly to the surface of the cerebral cortex. Therefore, the direction of the neuronal current flowing along the long apical dendrites of cortical pyramidal neurons also becomes perpendicular to the cortical surface [10, 22]. This physiological basis can be used as an important constraint for EEG source imaging [1], which will be introduced in Chap. 5.

There are two different sorts of intracellular potentials that may potentially contribute to the generation of scalp EEG signals, which are an *action potential* and a *postsynaptic potential*. The action potential is elicited by sudden changes in transmembrane resting potential due to the dynamic movements of intracellular and extracellular ions, such as sodium, chloride and potassium ions. When the action potential within a neuron propagates to a synapse, a small gap junction between two neurons, the postsynaptic potential is generated across a pair of neighboring neuronal membranes. If the postsynaptic potential exceeds a threshold level, the action potential of one neuron is delivered to the other neuron (see Fig. 1.1).

Among the two different types of potentials, the postsynaptic potential is believed to contribute more to the generation of measurable extracranial electric fields than the action potential. This is because the action potentials do not fire synchronously in a large number of neurons [25]. On the contrary, although the magnitude of the postsynaptic potential is generally smaller than that of the action potential, its relatively longer duration (~30 ms) enables synchronous generation of the postsynaptic potentials in a large number of neurons (see Fig. 1.1). As aforementioned, since the apical dendrites of cortical pyramidal neurons are arranged almost perpendicularly to the cortical surface, the summation of the synchronously generated postsynaptic

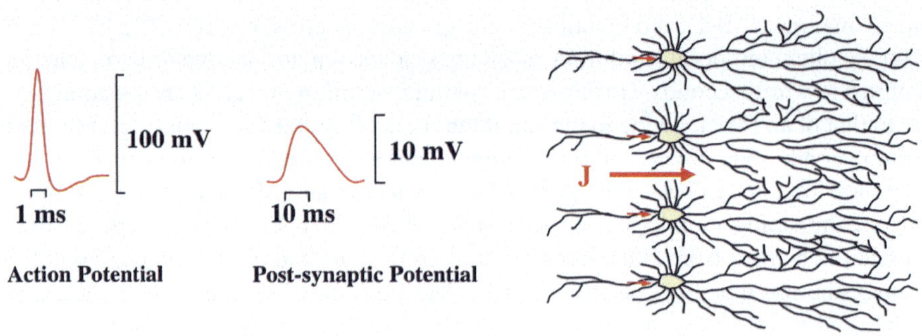

Fig. 1.1 (Left) Comparison of waveforms of action potential and postsynaptic potential. (Right) Synchronous occurrence of postsynaptic potentials can produce unidirectional primary current flow large enough to be recorded outside the head

potentials in a small cortical area can induce extracranial electric fields large enough to be measured on the scalp surface [1]. According to Hämäläinen et al. [10], the current density on the cortical surface is approximately 100 nA/mm^2. When numerous cortical neurons within a small area are activated synchronously, a unidirectional neuronal current flow is formed. Figure 1.1 depicts the comparison between action potential and postsynaptic potential, as well as a schematic illustration of the generation of the unidirectional neuronal current flow.

The unidirectional neuronal currents, which can be approximately modeled as *equivalent current dipoles* (ECDs) in EEG source imaging problems [6] (see Chap. 5 for more details), are called *primary* or *impressed currents* [22]. Since the human body is filled with electrically conductive media, the extracellular currents induced by the primary currents can flow even to the farthest part of the human body. These extracellular currents are known as *secondary*, *volume*, or *return currents* [22]. According to the electromagnetic theories, the flow of the secondary currents results in nonuniform potential distributions on the scalp. The measurement of the potential difference between two distant scalp locations over time is the EEG.

Because the EEG measures dynamic changes in potential differences originating from the secondary current flows, precise evaluation of conductivity profiles of the volume conductors, i.e., different tissue compartments inside the head, is important, not only to understand the underlying mechanisms of the EEG, but also to build a precise head model to calculate electric field quantities generated by primary neuronal currents (this process is called *forward calculation*). A human head can be roughly modelled with four different regions: brain, cerebrospinal fluid (CSF), skull, and scalp. Table 1.1 shows the typical conductivity values when the conductivity of each region is assumed to be isotropic (having uniform conductivity in all directions) and homogeneous [9]. The most notable point in the conductivity profile shown in Table 1.1 is that the conductivity value of the skull is even smaller than those of the other tissues. Because of the poor electrical conductivity of the skull, the secondary currents are severely distorted and/or attenuated before they are delivered to the scalp surface. Since the tissue conductivity is an important factor affecting the reliability and accuracy of EEG source imaging, anisotropic conductivity characteristics are sometimes considered. For example, the skull has an anisotropic conductivity property, approximately 0.014 and 0.0107 S/m for the directions normal and tangential to the skull surface, respectively [2]. White matter tissues also have an anisotropic conductivity property: the white matter conducts secondary currents much better along a fiber direction than in its transverse directions [31]. In practice, however, a rough approximation of the human head structure as piecewise isotropic and homogeneous volume conductors (e.g., brain, CSF, skull, and scalp) is most widely used. More detailed discussion of this topic is provided in Chap. 5.

Table 1.1 Typical conductivity values for different brain tissues/regions [9]

Regions	Absolute conductivity (S/m)	Relative conductivity
Brain	0.22	1
CSF	1.79	8
Skull	0.014	1/16
Scalp	0.22	1

1.2 Acquisition of EEG

Initial analog EEG devices recorded ongoing EEG activities on printed paper, when no quantitative EEG analysis was possible. Nowadays, owing to the development of computer technology and digital engineering, EEG signals are stored in computers as sampled numeric data. The use of a *digital EEG* enables us to utilize a variety of computational EEG analysis technologies, such as time-frequency analysis, functional connectivity analysis, and source imaging.

To record EEG data, at least two electrodes must be used, because EEG measures the potential difference between two distant scalp locations. Recent EEG recording devices allow simultaneous recording of EEG signals from many scalp locations. There are two types of EEG recording methods: *bipolar* and *unipolar* methods. In the bipolar method, electrodes are all paired, and the potential differences between each pair of electrodes are recorded. In the unipolar (or monopolar) method, the potential differences between each electrode and a reference electrode are recorded. Theoretically, the reference electrode in unipolar recording can be positioned anywhere; however, because the distribution of potential difference on the scalp surface varies according to the location of the reference electrode, *average reference* is frequently used. *Average-referenced* potential of each electrode can be readily evaluated by subtracting the average of all electrodes from the potential difference of each electrode. Average reference is particularly useful in depicting spatial distributions of potentials on the scalp surface, usually referred to as *topography* or *topographic map*.

EEG electrodes are generally attached on the scalp according to international standard configurations represented by the international 10–20 system. In the 10–20 system, electrodes are placed at 10 and 20% fractions of the geodesic distances between a number of anatomical landmarks such as inion, nasion, and two preauricular points. Smaller subdivisions (e.g., the 10–5 system) are also used for the placement of more electrodes. Further information on the electrode systems and electrode naming can be found in Oostenveld and Praamstra [24] and other sources—e.g., Wikipedia, https://en.wikipedia.org/wiki/10–20_system_(EEG).

In general, most EEG recording devices are composed of a signal amplifier, analog filter, and analog-to-digital converter (ADC). Use of high-quality signal amplifiers is necessary to display and process EEG signals on the order of microvolts. Since the recorded EEG signals are usually contaminated by unwanted environmental and/or systemic noises, such as alternating current (AC) power noises, a variety of electronic circuits are implemented in the EEG amplifier to remove or reduce the noise. Analog filters can also be used to remove specific noise components and increase signal-to-noise ratio (SNR). High-pass and band-reject (*notch*) filters can be used optionally to reject low-frequency physiological noise (e.g., respiration artifact) and AC power noise, respectively. All EEG devices should include an analog low-pass filter with a cutoff frequency less than half of the sampling rate to prevent *aliasing*, unwanted distortion in the sampled EEG signal. This type of analog low-pass filter is generally referred to as the *anti-aliasing filter*. This will be dealt with in a more detailed manner in Chap. 3. ADC converts the amplified and filtered analog signals to digital EEG signals using sampling and encoding procedures [28].

1.3 Computational EEG Analysis

Once the digital EEG signals have been stored in storage media, a variety of forms of information characterizing the underlying brain activities can be extracted from the numeric data. In this book, four major computational EEG analysis methods are introduced: EEG spectral analysis (Chap. 3), event-related potential (ERP) analysis (Chap. 4), EEG source imaging (Chap. 5), and functional connectivity analysis (Chap. 6).

1.3.1 EEG Spectral Analysis

One of the main advantages of EEG over the other hemodynamics- or neurochemistry-based neuroimaging modalities, such as fMRI and PET, is its superior temporal resolution that makes it possible to investigate neuronal activities changing on the order of tens of milliseconds. Thanks to the high temporal resolution of EEG, a large amount of useful information can also be obtained from frequency domain (or *spectral*) analysis. It is well known that changes in the EEG power spectrum are directly or indirectly associated with a variety of ongoing brain activities, e.g., mu-band (8–12 Hz) event-related desynchronization (ERD) and beta-band (18–22 Hz) event-related synchronization (ERS) associated with motor execution [11] and alpha-band (8–13 Hz) ERD associated with visual encoding [16]. EEG spectral analysis can also provide useful biomarkers to help diagnose and characterize various psychiatric diseases and neurological disorders. For example, reduced frontal gamma-band (30–50 Hz activity may indicate declined cognitive function [3] and increased midline beta-band (13–30 Hz) activity may be an indicator of restless-leg syndrome [8, 14]. Spectral analysis can also be used to implement various types of brain–computer interfaces (BCIs) and neurofeedback systems [12].

1.3.2 Event-Related Potential Analysis

In the history of EEG, the most important advancement was the use of *stimulus-locked* averaging of event-related EEG. Using *event-related potentials* (ERP) analysis, one can observe spatiotemporal components of stimulus-locked brain electrical activities with reduced background noise. Examples of important ERP components include P300 [20], N170 [4], mismatch negativity (MMN) [17], and error-related negativity (ERN) [30], which have been widely used not only for cognitive/clinical neuroscience studies [21] but also for BCI applications [7]. A series of methods has recently been proposed to extract more precise spatiotemporal ERP waveforms with fewer repeated trials, and this will be introduced in a detailed manner in Chap. 4.

1.3.3 EEG Source Imaging

The limited spatial resolution of EEG can be substantially enhanced by performing *EEG source imaging*, or *electrical source imaging* (ESI), which estimates locations, directions, and/or distribution of EEG sources by solving mathematically defined problems called inverse problems [23]. To solve the inverse problems, a procedure for modeling the human head and calculating the relationship between EEG sources and scalp potentials is necessary. This procedure is generally referred to as *forward calculation* or *solving forward problems*. Because accurate forward calculation is important to obtain accurate inverse solutions, high-precision numerical methods, such as the boundary element method (BEM) and finite-element method (FEM), have been adopted. To solve the inverse problems, various algorithms and models have been proposed, each of which has its own advantages and drawbacks. Detailed descriptions of the methods for EEG forward/inverse problems can be found in Chap. 5.

1.3.4 Functional Connectivity Analysis

Traditional neuroscience studies focused on functional specification of brain areas; however, recent neuroimaging studies exhibited increased interest in the functional connectivity among different brain areas. EEG is especially useful to study functional connectivity between two recording sites (or brain areas after EEG source imaging) because of its high temporal resolution. There are different kinds of functional connectivity measures that have been actively applied to EEG analyses, such as coherence, phase-locking value (PLV), phase lag index (PLI), Granger's causality (GC), and partial directed coherence (PDC). Functional connectivity analysis has proved to be useful to characterize various psychiatric diseases. Indeed, several recent studies have shown disrupted or abnormal functional connectivity patterns in patients with psychiatric illnesses; examples include schizophrenia [27], mild cognitive impairment [26], and post-traumatic stress disorder [13]. In particular, functional connectivity analysis is useful to study epilepsy because epilepsy is thought to be one of the most representative *brain network disorders* [18]. Detailed descriptions of the functional connectivity measures can be found in Chap. 6.

1.4 Applications of EEG

In the early stage of development of EEG, visual inspection of EEG waveforms was the only way to use EEG in practical applications. Indeed, visual inspection of EEG waveforms is still useful in studying sleep and diagnosing some neurological disorders, such as epilepsy. Dissemination of digital EEGs expanded the application fields of EEGs from limited research and diagnostic applications to more-extensive applications, including cognitive neuroscience study, diagnosis of psychiatric diseases, neuromarketing, neuroergonomics, sports science, and human brain mapping. Recently, owing to the rapid development of digital engineering, EEGs can be applied to real-time applications, such as BCI and neurofeedback.

The use of EEG in practical applications has steadily increased and is expected to continue to increase. Indeed, EEG has many advantages over the other methods to study brain functions, as follows:

- EEG is perfectly noninvasive, without any exposure to radiation or high magnetic field
- EEG is economical
- EEG devices can be made small and portable
- EEG has high temporal resolution
- EEG devices do not generate any noise
- EEG can be recorded in an open environment
- EEG can be acquired without active response from subjects.

Traditionally, EEG data were acquired in laboratory or clinical environments, where there are high-end EEG recording devices with a large number of channels and well-motivated participants who have agreed to participate in experiments with long durations. Recently, however, the advancement of wireless technology and high-performance biosensors enabled the development of wearable EEG devices that are easy to wear and comfortable for long-term use, expediting the development of novel applications of EEG that do not necessarily require laboratory settings, e.g., monitoring the brain activity of healthy persons during daily life [5, 19, 29].

Despite the recent development of EEG technology, EEG still has some intrinsic limitations that need to be overcome, examples of which include low spatial resolution and low SNR. Therefore, development of new computational EEG analysis methods is still necessary to enhance the reliability and usability of EEG.

References

1. S. Baillet, J.C. Mosher, R.M. Leahy, Electromagnetic brain mapping. IEEE Signal Process. Mag. **18**, 14–30 (2001)
2. S. Baillet, J.J. Riera, G. Marin, J.F. Mangin, J. Aubert, L. Garnero, Evaluation of inverse methods and head models for EEG source localization using a human skull phantom. Phys. Med. Biol. **46**, 77–96 (2001)
3. E. Başar, A review of gamma oscillations in healthy subjects and in cognitive impairment. Int. J. Psychophysiol. **90**, 99–117 (2013)
4. V.C. Blau, U. Maurer, N. Tottenham, B.D. McCandliss, The face-specific N170 component is modulated by emotional facial expression. Behav. Brain Funct. **3**, 7 (2007)
5. W.D. Chang, H.S. Cha, K. Kim, C.H. Im, Detection of eye blink artifacts from single prefrontal channel electroencephalogram. Comput. Methods Programs Biomed. **124**, 19–30 (2016)
6. J.C. de Munck, B.W. van Dijk, H. Spekreijse, Mathematical dipoles are adequate to describe realistic generators of human brain activity. IEEE Trans. Biomed. Eng. **35**(11), 960–966 (1988)
7. R. Fazel-Rezai, B.Z. Allison, C. Guger, E.W. Sellers, S.C. Kleih, A. Kübler, P300 brain computer interface: current challenges and emerging trends. Front Neuroeng. **5**, 14 (2012)
8. D.C. Hammond, Neurofeedback treatment of restless legs syndrome and periodic leg movements in sleep. J. Neurother. **16**(2), 155–163 (2012)
9. J. Haueisen, C. Ramon, M. Eiselt, H. Brauer, H. Nowak, Influence of tissue resistivities on neuromagnetic fields and electric potentials studied with a finite element model of the head. IEEE Trans. Biomed. Eng. **44**(8), 727–735 (1997)
10. M.S. Hämäläinen, R. Hari, R.J. Ilmoniemi, J. Knuutila, O.V. Lounasmaa, Magnetoencephalography. Theory, instrumentation and applications to the noninvasive study of human brain function. Rev. Mod. Phys. **65**, 413–497 (1993)
11. H.J. Hwang, K. Kwon, C.H. Im, Neurofeedback-based motor imagery training for brain-computer interface (BCI). J. Neurosci. Meth. **179**(1), 150–156 (2009)
12. H.J. Hwang, S. Kim, S. Choi, C.H. Im, EEG-based brain-computer interfaces: a thorough literature survey. Int. J. Hum. Comput. Interact. **29**(12), 814–826 (2013)
13. C. Imperatori, B. Farina, M.I. Quintiliani, A. Onofri, P. Castelli Gattinara, M. Lepore, V. Gnoni, E. Mazzucchi, A. Contardi, G. Della Marca, Aberrant EEG functional connectivity and EEG power spectra in resting state post-traumatic stress disorder: a sLORETA study. Biol. Psychol. **102**, 10–17 (2014)
14. K.Y. Jung, Y.S. Koo, B.J. Kim, D. Ko, G.T. Lee, K.H. Kim, C.H. Im, Electrophysiologic disturbances during daytime in patients with restless legs syndrome: further evidence of cognitive dysfunction? Sleep Med. **12**(4), 416–421 (2011)
15. Y.J. Jung, H.C. Kang, K.O. Choi, J.S. Lee, D.S. Kim, J.H. Cho, S.H. Kim, C.H. Im, H.D. Kim, Localization of ictal onset zones in Lennox-Gastaut syndrome using directional connectivity analysis of intracranial electroencephalography. Seizure **20**(6), 449–457 (2011)
16. W. Klimesch, R. Fellinger, R. Freunberger, Alpha oscillations and early stages of visual encoding. Front. Psychol. **2**, 118 (2011)
17. D. Ko, S. Kwon, G.T. Lee, C.H. Im, K.H. Kim, K.Y. Jung, Theta oscillation related to the auditory discrimination process in mismatch negativity: oddball versus control paradigm. J. Clin. Neurol. **8**(1), 35–42 (2012)
18. C. Lee, S.M. Kim, Y.J. Jung, C.H. Im, D.W. Kim, K.Y. Jung, Causal influence of epileptic network during spike-and-wave discharge in juvenile myoclonic epilepsy. Epilepsy Res. **108**(2), 257–266 (2014)
19. C.T. Lin, L.D. Liao, Y.H. Liu, I.J. Wang, B.S. Lin, J.Y. Chang, Novel dry polymer foam electrodes for long-term EEG measurement. IEEE Trans. Biomed. Eng. **58**, 1200–1207 (2011)
20. D.E. Linden, The p300: where in the brain is it produced and what does it tell us? Neuroscientist **11**(6), 563–576 (2005)
21. S.J. Luck, *An Introduction to the Event-Related Potential Technique* (The MIT Press, Boston, 2005)

22. J. Malmivuo, R. Plonsey, *Bioelectromagnetism* (Oxford University Press, Oxford, 1995)
23. P.L. Nunez, R. Srinivasan, *Electric Fields of the Brain: The Neurophysics of EEG* (Oxford University Press, Oxford, 2006)
24. R. Oostenveld, P. Praamstra, The five percent electrode system for high-resolution EEG and ERP measurements. Clin. Neurophysiol. **112**, 713–719 (2001)
25. W.W. Orrison, J.D. Lewine, J.A. Sanders, M.F. Hartshorne, *Functional Brain Imaging* (Mosby, Maryland Heights, 1995)
26. Y.A. Pijnenburg, Y. Vd Made Y, A.M. van Walsum, D.L. Knol, P. Scheltens, C.J. Stam, EEG synchronization likelihood in mild cognitive impairment and Alzheimer's disease during a working memory task. Clin. Neurophysiol. **15**(6), 1332–1339 (2004)
27. M. Shim, D.W. Kim, S.H. Lee, C.H. Im, Disruptions in small-world cortical functional connectivity network during an auditory oddball paradigm task in patients with schizophrenia. Schizophr. Res. **156**(2), 197–203 (2014)
28. M. Teplan, Fundamentals of EEG measurement. Meas. Sci. Rev. **2**(2), 1–11 (2002)
29. Y. Yasui, A brainwave signal measurement and data processing technique for daily life applications. J. Physiol. Anthropol. **28**, 145–150 (2009)
30. J.R. Wessel, Error awareness and the error-related negativity: evaluating the first decade of evidence. Front. Hum. Neurosci. **6**, 88 (2012)
31. C.H. Wolters, A. Anwander, X. Tricoche, D. Weinstein, M.A. Koch, R.S. MacLeod, Influence of tissue conductivity anisotropy on EEG/MEG field and return current computation in a realistic head model: a simulation and visualization study using high-resolution finite element modeling. Neuroimage **30**(3), 813–826 (2006)

Part II
Methods

Chapter 2
Preprocessing of EEG

Sung-Phil Kim

Abstract Preprocessing of the EEG signal, which is virtually a set of signal processing steps preceding main EEG data analyses, is essential to obtain only brain activity from the noisy EEG recordings. It has been shown that the design of preprocessing procedures can affect subsequent EEG data analysis outcomes. Preprocessing of EEG largely includes a number of processes, such as line noise removal, adjustment of referencing, elimination of bad EEG channels, and artifact removal. This chapter presents an overview of the methods available for each process and discusses practical considerations for applying these methods to the EEG signals. In particular, considerable attention is paid to the state-of-the-art artifact removal methods since there are still plenty of opportunities to enhance the artifact removal techniques for EEG, in the perspectives of both signal processing and neuroscience. It is desirable that this chapter provides the readers an overall view of EEG preprocessing pipelines and serves as a handbook guide for the practice of EEG preprocessing.

2.1 Introduction

Preprocessing of the EEG signal is an indispensable step for the analysis of EEG in most circumstances. Although there is still a lack of the standard pipeline of EEG preprocessing [8, 37, 58] it generally includes any necessary digital signal processing operations to polish up raw EEG signals with an aim to leave only brain activity signals for subsequent analyses. Often, EEG preprocessing also involves procedures to enhance spatiotemporal characteristics of the EEG signal related to the task used in a study [65].

A number of studies have demonstrated the influences of EEG preprocessing on the subsequent data analysis results [8, 33, 90, 110, 112]. For instance, the classification of different mental states from EEG or the control performance of a brain-

S.-P. Kim (✉)
School of Design and Human Engineering, Ulsan National Institute of Science and Technology, Ulsan, Republic of Korea
e-mail: spkim@unist.ac.kr

C.-H. Im (ed.), *Computational EEG Analysis*, Biological and Medical Physics, Biomedical Engineering, https://doi.org/10.1007/978-981-13-0908-3_2

computer interface (BCI) could be dependent on how EEG preprocessing treated the recorded EEG signals. In fact, it is obvious that any analytic result from the EEG signals containing significant noise and artifacts is likely to draw misleading conclusions. Recent reports also emphasize the standardization of preprocessing routines for multi-site data collection in divergent experimental environments [8, 37].

At the center of EEG processing lies the removal of any unnecessary covert and overt components of the EEG signals. In this chapter, we denote such unnecessary components as noise and artifacts. Following the previous notion [65], noise is regarded as neurological activities irrelevant to an examined behavioral task whereas artifacts are regarded to originate from external sources unrelated to neurological activities, such as eye movements, respiration or electrical interference. As most EEG preprocessing techniques pay attention to removing artefacts, we will also narrow our focus on the methods used to eliminate artifacts to clean up the EEG signals. Note that the topics covered by this chapter do not include the extraction of features from the EEG signals for particular applications, which should be discussed separately.

This chapter begins with the description of early-stage procedures to remove basic artifacts, sort out contaminated channels and possibly adjust references. It then discusses a range of methods to remove artifacts from the EEG signals, followed by brief discussion on EEG preprocessing.

2.2 Early-Stage Preprocessing

Early-stage EEG preprocessing involves fundamental and semi-automated organization of signal processing functions. It is distinguished from common artifact removal procedures as this stage of preprocessing is largely independent of any specific artifact. This chapter describes key parts of early-stage preprocessing including the removal of line noise, referencing and the elimination of bad channels. Before describing them, however, it is worth reviewing background characteristics of the EEG signals.

2.2.1 Characteristics of Background EEG

A basic and brief summary of the characteristics of background EEG activity is given as follows [104]. The frequency range of EEG is reportedly limited approximately from 0.01 to 100 Hz. The amplitudes of EEG generated from the brain typically range within $\pm 100\ \mu V$. The power spectral density of EEG is known to follow the power law [44]. Background brain rhythms are present in EEG, generally being classified in terms of oscillatory frequency into five disjoint bands: delta (0.5–4 Hz), theta (4–8 Hz), alpha (8–13 Hz), beta (13–30 Hz) and gamma (30–100 Hz). More details of the implications and functions of these rhythms can be found in other resources (e.g. see [10, 41, 63, 98]).

It is reasonable to consider the EEG signal as stochastic due to the lack of genuine EEG measurements [93]. In addition, over a long-term period, the EEG signals should be viewed as a non-stationary time series [57, 66]. However, EEG within a short time window can be approximately stationary with static statistical properties. The length of such a window containing stationary EEG signals varies with environments, generally ranging from several seconds to minutes [51].

2.2.2 *Line Noise Removal*

Most efforts to eliminate line noise from the EEG signal rely on notch filtering at 60 Hz. A notch filter is typically implemented with a certain frequency width surrounding 60 Hz (e.g. a width of 10 Hz). Consequently, notch filtering, although successfully removing line noise, could cause unintended distortions in signal components oscillating between 50 and 70 Hz. Also, the notch filter can reportedly generate a transient oscillation in baseline activity, leading to a potential issue in data interpretation [18]. Follow-up low-pass filtering with a cutoff frequency lower than 50 Hz may remedy this problem, but instead give rise to other issues such as alteration of temporal structures of EEG [106] or spurious interactions between EEG channels [40].

One suggestion to overcome this problem is estimating line noise embedded in the recorded EEG signals as precise as possible and subtracting it from the data [8, 80]. This method employs multi-taper decomposition to find line noise components in the signal. A short-time window slides over the course of the signal in which the transformation of EEG time series based on multi-tapers is carried out [5]. This transformation can effectively estimate spectral energy within each frequency band. Then, a regression model is applied to estimate the amplitude and phase of sinusoidal line noise (e.g. sinusoids at 60 Hz) in the transformed frequency domain. The Thompson F-test evaluates a significance of the magnitude of the estimated line noise. A time series of sinusoidal line noise is reconstructed if the magnitude is significant. This process is repeated over the sliding windows. The reconstructed line noise signal is subtracted from the original EEG signal. The entire process is repeated until the magnitude at the frequency of line noise becomes non-significant (Fig. 2.1). In this way, line noise components can be removed without damaging background spectral components [83].

2.2.3 *Referencing*

We often subtract a reference (with the same time resolution as the recorded EEG signals) from the original EEG signal at each channel. The reference signal should

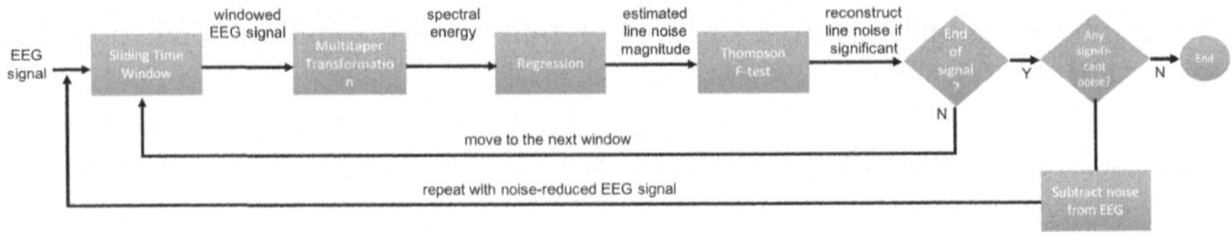

Fig. 2.1 Line noise removal using the multitaper transformation

remain unchanged relative to the EEG signals during the recording such that differences of the EEG signals from reference can effectively represent brain activity related to a study. Typical choices of reference include a signal recorded at a mastoid channel, an EEG signal at a particular channel, the average of two mastoid signals or the average of the entire EEG channels. In any case, it is strongly recommended that a researcher should inspect a chosen reference signal carefully to ensure that its amplitude level is on par with those of other EEG signals and it has no correlation with task-induced brain activity.

Referencing to a mastoid channel has a potential problem because it generates a single point of failure. If the contact to a mastoid becomes poor at any point during the recording, referencing to the mastoid can increase signal variance tremendously, resulting in irreversible contamination of EEG data. The same problem exists for referencing to a particular EEG channel. Using the common average reference (CAR) may reduce the effect of single-point failure [9], but still suffer from an outlier channel. One simple solution to this problem is detecting and removing bad channels before using CAR [8]. There are other systematic re-referencing methods developed to address the issues of reference, based on physical considerations and electrodynamics [38, 113, 114] or on statistical approaches [48, 69, 73].

2.2.4 Bad Channel Detection

It is often necessary to detect a noisy or *bad* channel that exhibits a contaminated EEG signal [8]. To detect a bad channel, we can screen each channel to identify EEG signals with excessively large amplitudes. The robust z-score can be used to detect extreme amplitudes. For instance, a bad channel is determined when it shows a robust z-score of the standard deviation greater than a threshold. A bad channel can be also detected by investigating correlation of a single channel with others. Normal EEG recordings show across-channel correlations in the low-frequency components. Hence, the correlation of one channel with other channels after low-pass filtering can allow us to detect bad channels. If two bad channels are incidentally correlated with each other, we can attempt to predict one channel using other channels. The predictor channels can be randomly selected from the remaining channels. Often, a contaminated channel exhibits relatively large energy in high-frequency bands. Thus, we can measure a ratio of the power of high-frequency components to that of low-frequency components and detect a bad channel showing a ratio higher than a threshold.

Once being detected, bad channels are replaced with virtual healthy channels created by the interpolation from neighboring channels, in order to reconstruct the global brain responses [8, 31]. There exist a number of interpolation schemes useful for channel reconstruction, including spherical splines [87], higher-order polynomials [4], nearest-neighbor averaging [15] and radial basis function [53]. Using spherical splines allows accurate estimation of scalp potentials if the electrode mapping is sufficiently dense [38, 97]. Interpolation using a statistical method such as radial basis functions has advantages of cost-effectiveness with less computational loads.

2.3 Artifact Removal

In this section, we briefly review the potential sources of artifacts mixed in the EEG signal and the techniques to remove or reduce artifacts. We primarily deal with artifact removal techniques, forgoing other steps of artifact management such as artifact detection. However, it does not mean that other methods including artifact detection or artifact avoidance are less crucial than artifact removal. In fact, artifact removal is often accompanied by artifact detection for efficient processing of artifacts. There have been a number of methods for artifact detection that the interested readers can refer to [3, 14, 32, 52, 81, 84].

2.3.1 Sources of Artifacts

The sources of EEG artifacts can be categorized into two classes: internal and external sources. The internal sources originate from the physiological systems of self and include electromagnetic activities of heart, eyes, muscle and so on. The external sources include all other possible signals from environments that can contaminate EEG such as wireless telecommunication signals, electrode attachment, recording equipment and cable movements [93]. Recently. the handling of external artifacts has become more important as EEG applications tend to move out of laboratories toward in-home healthcare systems [100]. Yet, the external sources, owing to their origins, can be inhibited once being identified. On the other hand, the internal artifacts physiologically permeate EEG, making it difficult to prevent them from occurring in advance. Therefore, most artifact removal methods have been focused on dealing with the internal artifacts and here we also pay our attention to the most pronounced internal artifacts that have been handled by EEG artifact removal methods.

Ocular artifacts include electric activities generated by eye movements or eye blinking [22, 23]. Interference by ocular artifacts is strong enough to be visible in EEG waveforms. EEG channels proximal to eyes are more vulnerable to ocular artifacts. Ocular artifacts can be detected by electrooculogram (EOG) measurements.

EOG recorded simultaneously with EEG offers an opportunity to readily remove ocular artifacts from EEG as it helps identify true profiles of artifacts. Once knowing the waveforms of ocular artifacts, removal algorithms can be developed to subtract them from the EEG signal without a need to reject contaminated EEG segments. To measure EOG for ocular artifact removal, it is recommended to record vertical (vEOG), horizontal (hEOG) and radial (rEOG) oculomotor signals [88].

Muscle artifacts include electric activities originating from muscle contraction of the body parts, including face, head, neck, limbs and others. Compared to ocular artifacts, muscle artifacts generate more various forms depending of the sources of muscles and related movements. The electrical signals associated with muscle artifacts can be measured by electromyogram (EMG). However, widespread sources of muscle artifacts over the body make it challenging to identify true profiles of artifacts. In addition, the spectral properties of cranial muscle artifacts vary across sources, corrupting high-frequency EEG components as well as low-frequency ones [93, 105]. The spatial distribution of muscle artifacts is wider than ocular artifacts, almost uniform over the entire scalp [44]. Temporal patterns of muscle artifacts are often associated with tasks as movements of subjects naturally occur in response to task requirements [95]. Considering all these issues, it still remains a significant challenge to remove muscle artifacts from EEG [76, 77, 95].

Cardiac artifacts originate from electric activities of the heart. Cardiac artifacts generally show low amplitudes compared to other artifacts. Cardiac electric activity can be measured by electrocardiography (ECG). They have well-known regular characteristics, which resemble epileptic EEG activity and thus possibly leading to incorrect seizure diagnosis [30]. However, for the perspective of removal algorithms, regular cardiac waveforms make it easier to correct in EEG. When an EEG electrode is positioned over a scalp artery, its contact with the skin can alter periodically due to recurrent motion of a pulsating vessel, which is likely to rhythmic electric activity similar to EEG oscillations [68]. But this pulsation effect shows periodicity synchronous with the heart, rendering itself being identified by ECG.

2.3.2 *Artifact Removal Methods*

Artifact removal methods aim to cancel or correct artifacts in EEG with minimal distortions in the brain signal. Here we briefly overview the computational methods to remove artifacts from EEG [52, 104]. Along this path, we avoid describing the details of mathematical backgrounds underlying each method (e.g. blind source separation (BSS), regression, linear transformation of multivariate Gaussian, etc.). Overall, an EEG artifact removal method belongs to one of the two kinds: a group of methods that corrects a single channel independently or another group that processes the whole channels all together. The single-channel processing methods employ various techniques including linear regression, filtering, wavelet transform and empirical mode decomposition (EMD). The whole-channel processing methods are based on BSS to estimate a set of hidden sources from an observed mixture of those sources with only limited information. Below we present several basic methods from both groups that have been most widely used in EEG studies.

2.3.2.1 Linear Regression

Assuming that artifact reference channels are available and contain thorough waveforms of artifacts, linear regression has been one of the main vehicles used to cancel artifacts from the EEG signal due to its simplicity and ease-of-use. A basic procedure is to estimate a portion of EEG contaminated by artifacts using regression and to subtract the regressed portion from the contaminated EEG [22, 23, 45]. Linear regression assumes that an EEG signal is the sum of an original brain signal and a fraction of the artifact represented in reference. It estimates this fractional factor from both the observed EEG signal and reference channel. The major drawbacks of linear regression are that one or more reference channels must be available (e.g. EOG or ECG), that it assumes a linear combination of EEG and artifacts where the EEG signal may possess internal nonlinear dynamics and non-stationary, and that it only applies well to a few types of artifacts such as EOG and ECG. However, if reference channels are available, linear regression is still an effective solution to remove artifacts [36, 107].

Linear regression methods operate particularly well with ocular artifacts since EOG can be directly measured or indirectly inferred from EEG [13, 42]. However, simple subtraction of a regressed portion of ocular artifacts from EEG can also take out cerebral components. This problem is termed bidirectional contamination [91]. Many methods have been proposed to address bidirectional contamination among which the aligned-artifact average procedure demonstrates promising results of canceling artifacts from eye movements or blinks while minimizing EEG contamination [21–23].

2.3.2.2 Filtering

Filters used for artifact removal build a statistical machine whose parameters are adaptively estimated with certain objectives, learning rules, model structures as well as data. Three types of filters have been primarily adopted for EEG artifact removal [104].

Adaptive filters model the way artifacts contaminate the EEG signal by adjusting the filter weights according to a learning rule formed by an optimization algorithm [47]. They assume no correlation between the EEG signal and artifacts. For example, let $x[n]$ be an observed EEG signal mixed with an unknown clean EEG signal $y[n]$ and an additive artifact signal $z[n]$ (i.e. $x[n] = y[n] + z[n]$). If the reference to artifact, $r[n]$, is available, the adaptive filter adjusts its weights, w, to minimize error between $x[n]$ and $w^T r[n]$. Since $r[n]$ is assumed to be uncorrelated with $y[n]$, the optimal weights would make $w^T r[n]$ as close to $z[n]$ as possible. Then, a difference, $\{x[n] - w^T r[n]\}$ will become close to $y[n]$ (Fig. 2.2). Many learning algorithms are available

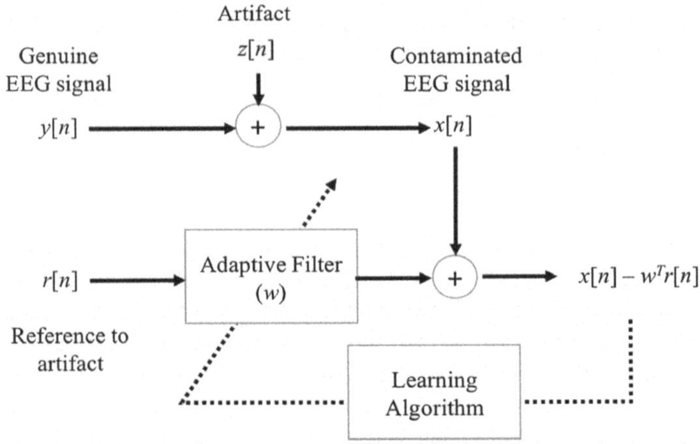

Fig. 2.2 EEG denoising with adaptive filtering and reference to artifacts

to adjust weights, including least mean squares (LMS) and recursive least squares (RLS) [47]. It has been shown that adaptive filters are superior to linear regression because proportion factors are less constrained [91]. However, as in linear regression, adaptive filters still require reference channels.

The Wiener filter is a linear time-invariant (LTI) filter that minimizes the mean squared error between desired response and filter output [47]. Optimal weights of the filter are estimated based on the Wiener-Hopf equation. Learning the weights is done offline with training samples that contain EEG and artifact signals. Having learned its weights, the Wiener filter can operate with the contaminated EEG signals without reference. However, the Wiener filter performance may deteriorate over time if a proportion of EEG contaminated by artifacts changes over time (i.e. non-stationary).

Bayesian filters in a linear or nonlinear form can overcome some shortcomings of both linear regression and the Wiener filter as they can sequentially update the states online without the need of reference channels. Here the states approximate unknown clean EEG signals. The system model in Bayesian filters approximates the sequential transition of clean EEG data according to the first-order Markov process and the observation model estimates the posterior probability distribution of clean EEG data after observing contaminated EEG data using a likelihood model and Bayesian approximation. The parameters of the system and observation models need to be learned from the training data as in the case of the Wiener filter. Although it is computationally expensive to estimate probability distributions in general, with some assumptions, Bayesian filters can reduce to simpler forms such as the Kalman filter or the particle filter. In particular, the Kalman filter has been widely applied for artifact removal for EEG [50, 59, 82].

2.3.2.3 Wavelet Transform and Empirical Mode Decomposition

EEG denoising can be achieved by decomposing a single-channel EEG signal into a set of fundamental basis signals, with a premise that some basis signals may contain the information of artifacts only. As such, we can find those artifact-related basis signals and remove them from the decomposed set. Two representative methods for decomposition of an EEG signal are presented below.

Wavelet transform convolves a given signal with a scaled and shifted version of a mother wavelet function. It results in a set of coefficients corresponding to each scale and time shift. The coefficients represent a similarity between a segment of the signal and the mother wavelet at a given scale. The discrete wavelet transform (DWT) is derived from continuous wavelet transform with discrete-time sampling. A basic procedure of the DWT is filtering a signal with low- and high-pass filters, respectively, where the low-pass filter works similar to the scaling function and the high-pass filter works similar to the mother wavelet function [52]. Then, the low-pass filtered output is passed to the next level of filtering with low- and high-pass filters again. This procedure is repeated up to K levels and yields one approximation coefficient and K detail coefficients where the approximation coefficient is obtained from the final low-pass filtering and the detail coefficients are obtained from a series of the high-pass filtering through K levels. Then, for denoising, a threshold is applied to the detail coefficients to sort out the ones with small magnitudes. It draws upon a hypothesis that the signal can be strongly correlated with a properly chosen mother wavelet basis at some levels whereas artifacts cannot be [104]. Finally, the artifact-reduced signal is reconstructed by the refined detail coefficients and the approximation coefficient [94]. Systematic ways of selecting a threshold can be found in some studies [34].

Empirical model decomposition (EMD) is a data-driven technique that decomposes a signal into a sum of the band-limited basis functions, called intrinsic mode functions (IMFs) [49]. The IMFs have zero means and are amplitude and frequency modulated. EMD has been shown to perform well with nonlinear and non-stationary signals. If different sets of IMFs can separately represent the signal and artifacts, we can reconstruct a clean EEG signal by removing artifact-related IMFs from the decomposed set. EMD has been successfully applied to artifact removal of EEG [70, 94, 115]. More advanced methods to overcome shortcomings of EMD (e.g. low robustness against noise, no mathematical background), including ensemble EMD (EEMD) [99, 116] and multivariate EMD (MEMD) [108], have also been adopted for artifact removal.

2.3.2.4 Blind Source Separation

Blind source separation (BSS) has been most widely used for artifact removal when the information about artifacts is limited—for instance, no reference is provided. The basic BSS methods used for artifact removal assume a linear mixture model in which the observed multi-channel EEG signals are assumed to be a linear mixture of unknown sources with little knowledge about sources or a mixing matrix. The optimal estimate of sources and a mixing matrix, thus, is achieved by certain assumptions on the sources such that the sources are mutually independent or uncorrelated. For instance, let **x** be an observed EEG signal vector, which is a mixture of an unknown source vector **s** with a mixing matrix A, given by:

$$\mathbf{x} = A\mathbf{s} + \mathbf{n} \tag{2.1}$$

where **n** denotes additive white noise.

Then, BSS methods estimate A to make sources in **s** as independent as possible. Once the estimate of A is obtained, its inverse matrix $W = A^{-1}$ is used to find the sources given by:

$$\mathbf{s} = W\mathbf{x}. \tag{2.2}$$

These estimated sources are then inspected either empirically (by visual inspection, for example) or automatically (by automatic source selection algorithms [109, 111, 119]) to identify artifact-related sources. The reduced set of sources after removing artifactual ones are then used to reconstruct artifact-free EEG data using A.

Despite its prevalence in EEG preprocessing, BSS suffers from limitations that it requires multi-channel EEG data and that there is always a possibility that removed sources may also carry information about brain activity. In addition, researchers should take into consideration the assumptions each BSS method works under, including independence, uncorrelatedness, and non-Gaussianity [54, 71]. A variety of BSS methods, however, have been successfully applied to remove artifacts from biomedical signals. Below are described several methods that have been widely used for EEG artifact removal.

Independent component analysis (ICA) is a BSS method based on assumptions of mutual linear independence between sources and non-Gaussianity [7]. ICA algorithms are based on either second-order or higher-order statistics [54]. The ICA algorithms based on higher-order statistics estimate W by maximizing statistical independence of the probability density functions of individual sources using mutual information or negentropy [7, 19]. The ICA algorithms based on second-order statistics estimate W by decorrelating the time-series data using the second-order blind identification (SOBI) [20, 103]. ICA has been reported to perform well in EEG artifact removal due to its reasonable assumption of statistical independence between the EEG signals and artifacts (e.g. see [2]). However, to explore statistical independence, ICA needs the sufficient amount of EEG data [56]. Also, ICA works best when the artifacts and the EEG signals remain stationary during the period of analysis, which may not be the case in general. To ensure stationarity, studies have suggested an epoch of 10 s or less, or a sample size in the order of multiples of $\sqrt{C}$ where C is the number of channels [56, 92]. When only a limited number of data samples are available, studies have suggested using the ICA algorithms with second-order statistics [28, 55].

Principal component analysis (PCA) has been proposed as a means to remove artifacts from EEG [39, 67, 102]. PCA transforms presumably correlated multichannel EEG data to mutually uncorrelated principal components (PCs) that preserve variance of the EEG data as much as possible. A set of PCs can represent artifacts if artifacts and brain signals are uncorrelated with each other. PCA also assumes joint normal distributions of the data. Often, it suffers from its restricted assumption that sources including brain activities are orthogonal to each other [39]. Hence, PCA is now seldom used directly for artifact removal but instead used for other essential preprocessing such as whitening [35].

Canonical correlation analysis (CCA) has also been extensively used for artifact removal from EEG [29, 43, 118]. Basically, CCA seeks for canonical variables that maximize correlations between two multivariate datasets. For EEG denoising, CCA finds canonical variables between the original data and its time-shifted version (typically one step behind). In doing so, canonical variables inferred in sequence represent the autocorrelation from the highest to the lowest. By assuming that brain activities are more correlated in time than artifacts, CCA identifies and removes canonical components with lower autocorrelations that may correspond to artifacts. The advantage of CCA over ICA is that it can take temporal correlations of the signals into account and use less computational resources [52].

Besides the three BSS methods described above, there are other BSS methods recently proposed for EEG artifact removal. Morphological component analysis (MCA) can decompose artifacts from EEG if the morphological template of the target artifacts is available [96]. Singular spectrum analysis (SSA) is a projective subspace method that projects a single-channel EEG signal onto a higher-dimensional space by time embedding, decomposes the embedded signal vector into uncorrelated components and reconstructs the EEG signal by projecting the embedded signals in the directions with large eigenvalues [24, 25, 101]. The sparse time artifact removal algorithm identifies and removes artifactual components of EEG that are sparse in both space and time [27].

2.3.2.5 Hybrid Artifact Removal Methods

Recent studies have proposed hybrid approaches for EEG artifact removal by combining more than one artifact removal algorithms. Many studies blend one algorithm from the BSS family and the other with decomposition (e.g. wavelet transform or EMD). A hybrid method can be characterized by the order of the applications of the selected algorithms. One group of methods first decomposes an EEG signal and then applies a BSS algorithm later whereas a different group of methods first estimates components using a BSS algorithm followed by a decomposition algorithm. The former usually corrects a single-channel EEG signal whereas the latter processes multi-channel EEG signals (Fig. 2.3). The hybrid approaches are generally designed to overcome the limitations of a single artifact removal approach and thus exhibit better performance, but require more careful choices of algorithms that fit adequately to the data and/or system requirements (e.g. computational complexity). The examples of the first group of hybrid methods for artifact removal, decomposition-BSS for single channels, can be found in various forms, applying wavelet transform followed

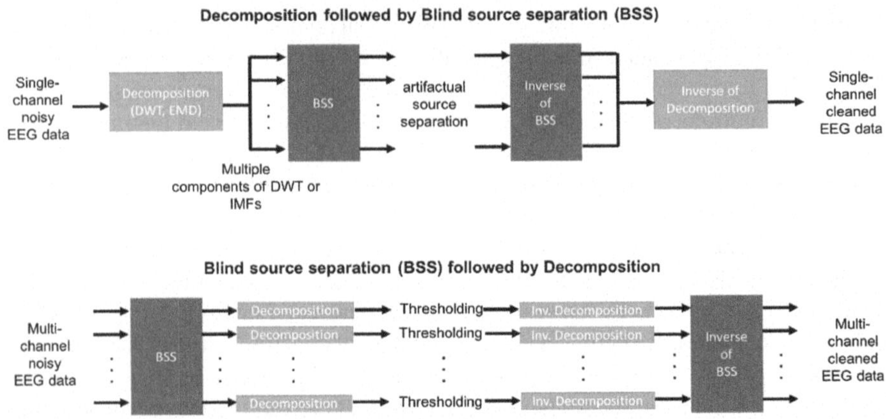

Fig. 2.3 Types of hybrid methods for EEG artifact removal

by (f.b.) ICA [11, 74, 75], EMD f.b. ICA [79, 117], and EMD f.b. CCA [16, 99]. The examples of the second group, BSS-decomposition for multiple channels, can also be found in different forms, including ICA f.b. wavelet [1, 12], stationary subspace analysis f.b. EMD [115], ICA f.b. EMD [70], ICA f.b. regression analysis [61], and ICA f.b. adaptive filtering [46].

2.4 Discussion

This chapter presents an overview of essential preprocessing steps for EEG. More detailed guidelines of practical preprocessing procedures can be found in existing literature (for instance, see [8, 52, 100, 104]). Although there has been substantial progress in the development of EEG preprocessing methods until recently, continuous advances in EEG-based research keep demanding innovations in preprocessing techniques. For instance, pervasive and ambulatory applications using EEG foster the development of preprocessing methods that can work with only a few channels in real time [78, 86]. Recent neuroscience approaches to use multi-modal brain measurements demand new ways of preprocessing EEG along with other signals such as functional magnetic resonance imaging (fMRI) [17]. EEG hyperscanning techniques recording brain activities simultaneously in more than one person, possibly over different sites, need a more systematic preprocessing procedure [6]. Here, we briefly discuss some ongoing issues and suggestions in the studies involving EEG preprocessing.

When comparing the artifact removal performance of different algorithms, often for the demonstration of the superiority of a newly proposed algorithm to existing ones, we can encounter the issue of the lack of ground truth. Since it is generally unknown about the exact waveform of a genuine EEG signal of interest, it is difficult to assess how much a noisy EEG signal become purified by an artifact removal algorithm [52]. One way to address this issue is to synthesize simulated signals mixed with putative true EEG signals and artifacts and evaluate an algorithm with the simulated signals [60, 64, 92]. Others have suggested using a well-known EEG waveform evoked by an established cognitive task to test artifact removal methods [104]. For example, an audio-visual task evoking the auditory N100 event-related potential may provide a validation dataset with which researchers can evaluate different artifact removal methods by assessing N100 waveforms after eliminating artifacts by different methods (see [88] for more details).

Besides performance evaluation discussed above, there are other issues to address for the development of an EEG artifact removal method. First, many recent EEG applications demand online preprocessing of artifacts [26, 43, 86]. Such online preprocessing is capable of detecting and removing artifacts even for non-stationary environments so that it can adaptively update the parameters of algorithms by tracking environmental changes. As such, the requirement of online processing sometimes weakens the advantages of certain algorithms that rely on the estimation of model parameters using a chunk of the training data (e.g. ICA or EMD). Also, computationally expensive machine learning algorithms (e.g. those with deep learning algorithms) may need further justification to be used for online processing. Yet, in the course of the development of a new artifact removal algorithm, it would be more effective to consider online implementation if possible. A fully automated artifact removal algorithm will underpin online implementation [26, 84]. Second, the availability of reference channels should be taken into consideration for artifact removal. If no reference channel is available, we need to use prior knowledge about artifacts or infer artifacts directly from EEG data [62, 72, 86]. Generally, using an explicit reference channel may help customizing algorithms for each individual, yielding a more precise preprocessing method. Depending on the types of artifacts, it may be useful for improving EEG preprocessing to utilize reference channels, often acquired with a separate device, such as: EOG channel [22, 23, 61], ECG channel [30], eye tracker [85], accelerometer [24, 25], and contact impedance [119]. Third, it would be crucial to match the properties of an algorithm with statistical and physiological characteristics of the artifacts to remove. The readers may refer to Urigüen et al. [104] for the suggestions of artifact removal algorithms suitable for different types of artifacts. Fourth, researchers often opt to utilize public software tools for EEG preprocessing as well as other EEG data analyses (see [52] for the list of available software tools). Even though a number of software tools offer complete preprocessing routines and user interfaces for EEG studies, it is recommended to intensively explore the theoretical backgrounds and technical details of a tool being used. Otherwise, it is difficult to understand how EEG signals are processed at each preprocessing step. Fifth, it is helpful to inform study participants about the problems of artifacts in EEG recordings such that participants can minimize their movements during the main tasks [89]. Although it would be also problematic if participants pay too much attention to movement restriction throughout the whole experiment, a short training phase for participants to minimize movements during the task periods interleaved with more flexible breaks can help acquiring high-quality EEG data at the stage of recording.

This instruction would be especially crucial for the studies recruiting younger participants. Sixth, not only highly contaminated channels but also highly contaminated trials are often eliminated from the analysis. The elimination of contaminated trials is usually conducted after all the preprocessing steps but its operational principle is similar to other preprocessing methods. Generally, the trials containing the EEG signal magnitude greater than a threshold level (e.g. $\pm150\,\mu$V) are classified as being contaminated [89]. Here, the threshold must be specified depending on experimental conditions. Rejection of too many trials would cause a shortage of the amount of data in the subsequent analyses, so a careful interactive investigation between preprocessing methods and trial rejection should be considered. Finally, a developed preprocessing pipeline may call for assessments based on feedbacks from the designated applications (e.g. classification of the user intention for brain-computer interfaces). Consequently, it is worth deliberating an *end-to-end* design of EEG signal processing, from the recording to the interpretation of EEG as a whole.

References

1. M.T. Akhtar, W. Mitsuhashi, C.J. James, Employing spatially constrained ICA and wavelet denoising, for automatic removal of artifacts from multichannel EEG data. Sig. Process. **92**, 401–416 (2012)
2. L. Albera, A. Kachenoura, P. Comon, A. Karfoul, F. Wendling, L. Senhadji, I. Merlet, ICA-based EEG denoising: a comparative analysis of fifteen methods. Bull. Pol. Ac: Tech. **60**, 407–418 (2012)
3. M.N. Anastasiadou, M. Christodoulakis, E.S. Papathanasiou, S.S. Papacostas, G.D. Mitsis, Unsupervised detection and removal of muscle artifacts from scalp EEG recordings using canonical correlation analysis, Wavelets and random forests. Clin. Neurophysiol. **128**, 1755–1769 (2017)
4. H. Ashida, J. Tatsuno, J. Okamoto, E. Maru, Field mapping of EEG by unbiased polynomial interpolation. Comput. Biomed. Res. **17**, 267–276 (1984)
5. B. Babadi, E.N. Brown, A review of multitaper spectral analysis. IEEE Trans. Biomed. Eng. **61**, 1555–1564 (2014)
6. F. Babiloni, L. Astolfi, Social neuroscience and hyperscanning techniques: past, present and future. Neurosci. Biobehav. Rev. **44**, 76–93 (2014)
7. A.J. Bell, T.J. Sejnowski, An information-maximization approach to blind separation and blind deconvolution. Neural Comput. **7**, 1129–1159 (1995)
8. N. Bigdely-Shamlo, T. Mullen, C. Kothe, K.-M. Su, K.A. Robbins, The PREP pipeline: standardized preprocessing for large-scale EEG analysis. Front. Neuroinform. **9**, 16 (2015)
9. C. Binnie, R. Cooper, F. Mauguiere, J. Osselton, P. Prior, B. Tedman, EEG, Pediatric neurophysiology, special techniques and applications, in *Clinical Neurophysiology*, vol. 2, ed. by R.M.F. Cooper, J.W. Osselton, P.F. Prior, B.M. Tedman (Elsevier, Amsterdam, 2003)
10. G. Buzsáki, X.-J. Wang, Mechanisms of gamma oscillations. Annu. Rev. Neurosci. **35**, 203–225 (2012)
11. S. Calcagno, F. La Foresta, M. Versaci, Independent component analysis and discrete wavelet transform for artifact removal in biomedical signal processing. Am. J. Appl. Sci. **11**, 57 (2014)
12. N.P. Castellanos, V.A. Makarov, Recovering EEG brain signals: artifact suppression with wavelet enhanced independent component analysis. J. Neurosci. Methods **158**, 300–312 (2006)

13. W.-D. Chang, H.-S. Cha, K. Kim, C.-H. Im, Detection of eye blink artifacts from single prefrontal channel electroencephalogram. Comput. Methods Programs Biomed. **124**, 19–30 (2016)
14. M. Chaumon, D.V. Bishop, N.A. Busch, A practical guide to the selection of independent components of the electroencephalogram for artifact correction. J. Neurosci. Methods **250**, 47–63 (2015)
15. N. Chauveau, J. Morucci, X. Franceries, P. Celsis, B. Rigaud, Resistor mesh model of a spherical head: Part 1: Applications to scalp potential interpolation. Med. Biol. Eng. Comput. **43**, 694–702 (2005)
16. X. Chen, C. He, H. Peng, Removal of muscle artifacts from single-channel EEG based on ensemble empirical mode decomposition and multiset canonical correlation analysis. J. Appl. Math. **2014** (2014)
17. M.E. Chowdhury, K.J. Mullinger, P. Glover, R. Bowtell, Reference layer artefact subtraction (RLAS): a novel method of minimizing EEG artefacts during simultaneous fMRI. NeuroImage **84**, 307–319 (2014)
18. J.L. Cole, G. Goldberg, Central nervous system electrophysiology, in *Physical Medicine & Rehabilitation: Principles and Practice*, 4th edn., ed. by J.A. DeLisa, B.M. Gans, N.E. Walsh (Lippincott Williams & Wilkins, Philadelphia, 2005)
19. P. Comon, Independent component analysis, a new concept? Sig. Process. **36**, 287–314 (1994)
20. M. Congedo, C. Gouy-Pailler, C. Jutten, On the blind source separation of human electroencephalogram by approximate joint diagonalization of second order statistics. Clin. Neurophysiol. **119**, 2677–2686 (2008)
21. R.J. Croft, R.J. Barry, EOG correction: a new aligned-artifact average solution. Clin. Neurophysiol. **107**, 395–401 (1998)
22. R.J. Croft, R.J. Barry, EOG correction of blinks with saccade coefficients: a test and revision of the aligned-artefact average solution. Clin. Neurophysiol. **111**, 444–451 (2000)
23. R.J. Croft, R.J. Barry, Removal of ocular artifact from the EEG: a review. Neurophysiol. Clin. **30**, 5–19 (2000)
24. I. Daly, M. Billinger, R. Scherer, G. Müller-Putz, On the automated removal of artifacts related to head movement from the EEG. IEEE Trans. Neural. Syst. Rehab. Eng. **21**, 427–434 (2013)
25. I. Daly, N. Nicolaou, S.J. Nasuto, K. Warwick, Automated artifact removal from the electroencephalogram: a comparative study. Clin. EEG Neurosci. **44**, 291–306 (2013)
26. I. Daly, R. Scherer, M. Billinger, G. Müller-Putz, FORCe: fully online and automated artifact removal for brain-computer interfacing. IEEE Trans. Neural. Syst. Rehab. Eng. **23**, 725–736 (2015)
27. A. de Cheveigné, Sparse time artifact removal. J. Neurosci. Methods **262**, 14–20 (2016)
28. A. de Cheveigné, L.C. Parra, Joint decorrelation, a versatile tool for multichannel data analysis. NeuroImage **98**, 487–505 (2014)
29. W. De Clercq, A. Vergult, B. Vanrumste, W. Van Paesschen, S. Van Huffel, Canonical correlation analysis applied to remove muscle artifacts from the electroencephalogram. IEEE Trans. Biomed. Eng. **53**, 2583–2587 (2006)
30. M. De Vos, W. Deburchgraeve, P. Cherian, V. Matic, R. Swarte, P. Govaert, G.H. Visser, S. Van Huffel, Automated artifact removal as preprocessing refines neonatal seizure detection. Clin. Neurophysiol. **122**, 2345–2354 (2011)
31. A. Delorme, S. Makeig, EEGLAB: an open source toolbox for analysis of single-trial EEG dynamics including independent component analysis. J. Neurosci. Methods **134**, 9–21 (2004)
32. A. Delorme, T. Sejnowski, S. Makeig, Enhanced detection of artifacts in EEG data using higher-order statistics and independent component analysis. NeuroImage **34**, 1443–1449 (2007)
33. A. Doering, H. Jäger, H. Witte, M. Galicki, C. Schelenz, M. Specht, K. Reinhart, M. Eiselt, Adaptable preprocessing units and neural classification for the segmentation of EEG signals. Meth. Inform. Med. **38**, 214–224 (1999)
34. D.L. Donoho, I.M. Johnstone, Adapting to unknown smoothness via wavelet shrinkage. J. Am. Stat. Assoc. **90**, 1200–1224 (1995)

35. D.A. Engemann, A. Gramfort, Automated model selection in covariance estimation and spatial whitening of MEG and EEG signals. NeuroImage **108**, 328–342 (2015)
36. I.D. Evans, G. Jamieson, R. Croft, T.T. Pham, Empirically validating fully automated EOG artifact correction using independent components analysis, in *Abstracts of the ACNS-2012 Australasian Cognitive Neuroscience Conference* (2012)
37. F. Farzan, S. Atluri, M. Frehlich, P. Dhami, K. Kleffner, R. Price, R.W. Lam, B.N. Frey, R. Milev, A. Ravindran, Standardization of electroencephalography for multi-site, multi-platform and multi-investigator studies: insights from the canadian biomarker integration network in depression. Sci. Rep. **7**, 7473 (2017)
38. T.C. Ferree, Spherical splines and average referencing in scalp electroencephalography. Brain Topogr. **19**, 43–52 (2006)
39. S.P. Fitzgibbon, D.M. Powers, K.J. Pope, C.R. Clark, Removal of EEG noise and artifact using blind source separation. J. Clin. Neurophysiol. **24**, 232–243 (2007)
40. E. Florin, J. Gross, J. Pfeifer, G.R. Fink, L. Timmermann, The effect of filtering on Granger causality based multivariate causality measures. NeuroImage **50**, 577–588 (2010)
41. W.J. Freeman, Mechanism and significance of global coherence in scalp EEG. Curr. Opin. Neurobiol. **31**, 199–205 (2015)
42. G. Gómez-Herrero, W. De Clercq, H. Anwar, O. Kara, K. Egiazarian, S. Van Huffel, W. Van Paesschen, Automatic removal of ocular artifacts in the EEG without an EOG reference channel, in *Abstracts of the Proceedings of the 7th Nordic* (2006)
43. J. Gao, C. Zheng, P. Wang, Online removal of muscle artifact from electroencephalogram signals based on canonical correlation analysis. Clin. EEG Neurosci. **41**, 53–59 (2010)
44. I.I. Goncharova, D.J. McFarland, T.M. Vaughan, J.R. Wolpaw, EMG contamination of EEG: spectral and topographical characteristics. Clin. Neurophysiol. **114**, 1580–1593 (2003)
45. G. Gratton, M.G. Coles, E. Donchin, A new method for off-line removal of ocular artifact. Electroencephalogr. Clin. Neurophysiol. **55**, 468–484 (1983)
46. C. Guerrero-Mosquera, A. Navia-Vázquez, Automatic removal of ocular artefacts using adaptive filtering and independent component analysis for electroencephalogram data. IET Signal Proc. **6**, 99–106 (2012)
47. S.O. Haykin, *Adaptive Filter Theory*, 5th ed. (Pearson Higher (ed.), Upper Saddle River, New Jersey, 2013)
48. S. Hu, M. Stead, G.A. Worrell, Automatic Identification and Removal of Scalp Reference Signal for Intracranial EEGs Based on Independent Component Analysis. IEEE Trans. Biomed. Eng. **54**, 1560–1572 (2007)
49. N.E. Huang, Z. Shen, S.R. Long, M.C. Wu, H.H. Shih, Q. Zheng, N.-C. Yen, C.C. Tung, H.H. Liu, The empirical mode decomposition and the Hilbert spectrum for nonlinear and non-stationary time series analysis, in *Abstracts of the Proceedings of the Royal Society of London A: Mathematical, Physical and Engineering Sciences* (1998)
50. M.H. In, S.Y. Lee, T.S. Park, T.-S. Kim, M.H. Cho, Y.B. Ahn, Ballistocardiogram artifact removal from EEG signals using adaptive filtering of EOG signals. Physiol. Meas. **27**, 1227 (2006)
51. T. Inouye, S. Toi, Y. Matsumoto, A new segmentation method of electroencephalograms by use of Akaike's information criterion. Cogn. Brain. Res. **3**, 33–40 (1995)
52. M.K. Islam, A. Rastegarnia, Z. Yang, Methods for artifact detection and removal from scalp EEG: a review. Neurophysiol. Clin./Clin. Neurophysiol. **46**, 287–305 (2016)
53. J. Jäger, A. Klein, M. Buhmann, W. Skrandies, Reconstruction of electroencephalographic data using radial basis functions. Clin. Neurophysiol. **127**, 1978–1983 (2016)
54. C.J. James, C.W. Hesse, Independent component analysis for biomedical signals. Physiol. Meas. **26**, R15 (2004)
55. C.A. Joyce, I.F. Gorodnitsky, M. Kutas, Automatic removal of eye movement and blink artifacts from EEG data using blind component separation. Psychophysiology **41**, 313–325 (2004)
56. T.-P. Jung, S. Makeig, C. Humphries, T.-W. Lee, M.J. Mckeown, V. Iragui, T.J. Sejnowski, Removing electroencephalographic artifacts by blind source separation. Psychophysiology **37**, 163–178 (2000)

57. J.P. Kaipio, P.A. Karjalainen, Simulation of nonstationary EEG. Biol. Cybern. **76**, 349–356 (1997)
58. A. Keil, S. Debener, G. Gratton, M. Junghöfer, E.S. Kappenman, S.J. Luck, P. Luu, G.A. Miller, C.M. Yee, Committee report: publication guidelines and recommendations for studies using electroencephalography and magnetoencephalography. Psychophysiology **51**, 1–21 (2014)
59. J.J. Kierkels, J. Riani, J.W. Bergmans, G.J. Van Boxtel, Using an eye tracker for accurate eye movement artifact correction. IEEE Trans. Biomed. Eng. **54**, 1256–1267 (2007)
60. J.J. Kierkels, G.J. van Boxtel, L.L. Vogten, A model-based objective evaluation of eye movement correction in EEG recordings. IEEE Trans. Biomed. Eng. **53**, 246–253 (2006)
61. M.A. Klados, C. Papadelis, C. Braun, P.D. Bamidis, REG-ICA: a hybrid methodology combining blind source separation and regression techniques for the rejection of ocular artifacts. Biomed. Signal Process. Control **6**, 291–300 (2011)
62. A. Klein, W. Skrandies, A reliable statistical method to detect eyeblink-artefacts from electroencephalogram data only. Brain Topogr. **26**, 558–568 (2013)
63. W. Klimesch, P. Sauseng, S. Hanslmayr, EEG alpha oscillations: the inhibition–timing hypothesis. Brain Res. Rev. **53**, 63–88 (2007)
64. P. Krishnaswamy, G. Bonmassar, C. Poulsen, E.T. Pierce, P.L. Purdon, E.N. Brown, Reference-free removal of EEG-fMRI ballistocardiogram artifacts with harmonic regression. NeuroImage **128**, 398–412 (2016)
65. D.J. Krusienski, D.J. McFarland, J.C. Principe, BCI signal processing: feature extraction, in *Brain-Computer Interfaces: Principles and Practice*, ed. by J.R. Wolpaw, E.W. Wolpaw (Oxford University Press, Oxford, 2012)
66. A.D. Krystal, R. Prado, M. West, New methods of time series analysis of non-stationary EEG data: Eigenstructure decompositions of time varying autoregressions. Clin. Neurophysiol. **110**, 2197–2206 (1999)
67. T.D. Lagerlund, F.W. Sharbrough, N.E. Busacker, Spatial filtering of multichannel electroencephalographic recordings through principal component analysis by singular value decomposition. J. Clin. Neurophysiol. **14**, 73–82 (1997)
68. J.-P. Lanquart, M. Dumont, P. Linkowski, QRS artifact elimination on full night sleep EEG. Med. Eng. Phys. **28**, 156–165 (2006)
69. K.Q. Lepage, M.A. Kramer, C.J. Chu, A statistically robust EEG re-referencing procedure to mitigate reference effect. J. Neurosci. Methods **235**, 101–116 (2014)
70. J.P. Lindsen, J. Bhattacharya, Correction of blink artifacts using independent component analysis and empirical mode decomposition. Psychophysiology **47**, 955–960 (2010)
71. G. Lio, P. Boulinguez, Greater robustness of second order statistics than higher order statistics algorithms to distortions of the mixing matrix in blind source separation of human EEG: implications for single-subject and group analyses. NeuroImage **67**, 137–152 (2013)
72. J. Ma, P. Tao, S. Bayram, V. Svetnik, Muscle artifacts in multichannel EEG: characteristics and reduction. Clin. Neurophysiol. **123**, 1676–1686 (2012)
73. N. Madhu, R. Ranta, L. Maillard, L. Koessler, A unified treatment of the reference estimation problem in depth EEG recordings. Med. Biol. Eng. Comput. **50**, 1003–1015 (2012)
74. N. Mammone, F. La Foresta, F.C. Morabito, Automatic artifact rejection from multichannel scalp EEG by wavelet ICA. IEEE Sens. J. **12**, 533–542 (2012)
75. N. Mammone, F.C. Morabito, Enhanced automatic wavelet independent component analysis for electroencephalographic artifact removal. Entropy **16**, 6553–6572 (2014)
76. B.W. McMenamin, A.J. Shackman, L.L. Greischar, R.J. Davidson, Electromyogenic artifacts and electroencephalographic inferences revisited. NeuroImage **54**, 4–9 (2011)
77. B.W. McMenamin, A.J. Shackman, J.S. Maxwell, D.R. Bachhuber, A.M. Koppenhaver, L.L. Greischar, R.J. Davidson, Validation of ICA-based myogenic artifact correction for scalp and source-localized EEG. NeuroImage **49**, 2416–2432 (2010)
78. V. Mihajlović, B. Grundlehner, R. Vullers, J. Penders, Wearable, wireless EEG solutions in daily life applications: what are we missing? IEEE J. Biomed. Health Inform. **19**, 6–21 (2015)
79. B. Mijovic, M. De Vos, I. Gligorijevic, J. Taelman, S. Van Huffel, Source separation from single-channel recordings by combining empirical-mode decomposition and independent component analysis. IEEE Trans. Biomed. Eng. **57**, 2188–2196 (2010)

80. P.P. Mitra, B. Pesaran, Analysis of dynamic brain imaging data. Biophys. J. **76**, 691–708 (1999)
81. A. Mognon, J. Jovicich, L. Bruzzone, M. Buiatti, ADJUST: an automatic EEG artifact detector based on the joint use of spatial and temporal features. Psychophysiology **48**, 229–240 (2011)
82. F. Morbidi, A. Garulli, D. Prattichizzo, C. Rizzo, P. Manganotti, S. Rossi, Off-line removal of TMS-induced artifacts on human electroencephalography by Kalman filter. J. Neurosci. Methods **162**, 293–302 (2007)
83. T. Mullen, NITRC: CleanLine: Tool/Resource Info (2012)
84. H. Nolan, R. Whelan, R. Reilly, FASTER: fully automated statistical thresholding for EEG artifact rejection. J. Neurosci. Methods **192**, 152–162 (2010)
85. B. Noureddin, P.D. Lawrence, G.E. Birch, Online removal of eye movement and blink EEG artifacts using a high-speed eye tracker. IEEE Trans. Biomed. Eng. **59**, 2103–2110 (2012)
86. H. Peng, B. Hu, Q. Shi, M. Ratcliffe, Q. Zhao, Y. Qi, G. Gao, Removal of ocular artifacts in EEG—An improved approach combining DWT and ANC for portable applications. IEEE J. Biomed. Health Inform. **17**, 600–607 (2013)
87. F. Perrin, J. Pernier, O. Bertnard, M. Giard, J. Echallier, Mapping of scalp potentials by surface spline interpolation. Electroencephalogr. Clin. Neurophysiol. **66**, 75–81 (1987)
88. T.T. Pham, R.J. Croft, P.J. Cadusch, R.J. Barry, A test of four EOG correction methods using an improved validation technique. Int. J. Psychophysiol. **79**, 203–210 (2011)
89. T. Picton, S. Bentin, P. Berg, E. Donchin, S. Hillyard, R. Johnson, G. Miller, W. Ritter, D. Ruchkin, M. Rugg, Guidelines for using human event-related potentials to study cognition: recording standards and publication criteria. Psychophysiology **37**, 127–152 (2000)
90. I. Rejer, P. Górski, Independent Component Analysis for EEG data preprocessing-algorithms comparison, in *Computer Information Systems and Industrial Management*, vol. 8104, ed. by K. Saeed, R. Chaki, A. Cortesi, S. Wierzchoń (Springer, Heidelberg, 2013)
91. S. Romero, M. Mañanas, M.J. Barbanoj, Ocular reduction in EEG signals based on adaptive filtering, regression and blind source separation. Ann. Biomed. Eng. **37**, 176–191 (2009)
92. S. Romero, M.A. Mañanas, M.J. Barbanoj, A comparative study of automatic techniques for ocular artifact reduction in spontaneous EEG signals based on clinical target variables: a simulation case. Comput. Biol. Med. **38**, 348–360 (2008)
93. L. Sörnmo, P. Laguna, *Bioelectrical Signal Processing in Cardiac and Neurological Applications* (Academic Press, Elsevier, Amsterdam, 2005)
94. D. Safieddine, A. Kachenoura, L. Albera, G. Birot, A. Karfoul, A. Pasnicu, A. Biraben, F. Wendling, L. Senhadji, I. Merlet, Removal of muscle artifact from EEG data: comparison between stochastic (ICA and CCA) and deterministic (EMD and wavelet-based) approaches. EURASIP J. Adv. Signal Process. **2012**, 127 (2012)
95. A.J. Shackman, B.W. McMenamin, H.A. Slagter, J.S. Maxwell, L.L. Greischar, R.J. Davidson, Electromyogenic artifacts and electroencephalographic inferences. Brain Topogr. **22**, 7–12 (2009)
96. B. Singh, H. Wagatsuma, A removal of eye movement and blink artifacts from EEG data using morphological component analysis. Comput. Math. Methods Med. (2017)
97. A.C. Soong, J.C. Lind, G.R. Shaw, Z.J. Koles, Systematic comparisons of interpolation techniques in topographic brain mapping. Clin. Neurophysiol. **87**, 185–195 (1993)
98. M. Steriade, Grouping of brain rhythms in corticothalamic systems. Neuroscience **137**, 1087–1106 (2006)
99. K.T. Sweeney, S.F. McLoone, T.E. Ward, The use of ensemble empirical mode decomposition with canonical correlation analysis as a novel artifact removal technique. IEEE Trans. Biomed. Eng. **60**, 97–105 (2013)
100. K.T. Sweeney, T.E. Ward, S.F. McLoone, Artifact removal in physiological signals—practices and possibilities. IEEE Trans. Inf. Technol. Biomed. **16**, 488–500 (2012)
101. A.R. Teixeira, A.M. Tomé, K. Stadlthanner, E.W. Lang, (eds.), Nonlinear projective techniques to extract artifacts in biomedical signals. Signal Processing Conference, 2006 14th European; IEEE (2006)

102. E.M. ter Braack, B. de Jonge, M.J. van Putten, Reduction of TMS induced artifacts in EEG using principal component analysis. IEEE Trans. Neural Syst. Rehab. Eng. **21**, 376–382 (2013)
103. K. Ting, P. Fung, C. Chang, F. Chan, Automatic correction of artifact from single-trial event-related potentials by blind source separation using second order statistics only. Med. Eng. Phys. **28**, 780–794 (2006)
104. J.A. Urigüen, B. Garcia-Zapirain, EEG artifact removal—state-of-the-art and guidelines. J. Neural Eng. **12**, 031001 (2015)
105. A. Van Boxtel, Optimal signal bandwidth for the recording of surface EMG activity of facial, jaw, oral, and neck muscles. Psychophysiology **38**, 22–34 (2001)
106. R. VanRullen, Four common conceptual fallacies in mapping the time course of recognition. Front. Psychol. **2**, 365 (2011)
107. G.L. Wallstrom, R.E. Kass, A. Miller, J.F. Cohn, N.A. Fox, Automatic correction of ocular artifacts in the EEG: a comparison of regression-based and component-based methods. Int. J. Psychophysiol. **53**, 105–119 (2004)
108. G. Wang, C. Teng, K. Li, Z. Zhang, X. Yan, The removal of EOG artifacts from EEG signals using independent component analysis and multivariate empirical mode decomposition. IEEE J. Biomed. Health Inform. **20**, 1301–1308 (2016)
109. I. Winkler, S. Brandl, F. Horn, E. Waldburger, C. Allefeld, M. Tangermann, Robust artifactual independent component classification for BCI practitioners. J. Neural Eng. **11**, 035013 (2014)
110. I. Winkler, S. Debener, K.-R. Müller, M. Tangermann, On the influence of high-pass filtering on ICA-based artifact reduction in EEG-ERP, in *Abstracts of the Engineering in Medicine and Biology Society, EMBC, 37th Annual International Conference of the IEEE* (2015)
111. I. Winkler, S. Haufe, M. Tangermann, Automatic classification of artifactual ICA-components for artifact removal in EEG signals. Behav. Brain Funct. **7**, 30 (2011)
112. D. Wu, J.-T. King, C.-H. Chuang, C.-T. Lin, T.-P. Jung, Spatial filtering for EEG-based regression problems in brain-computer interface (BCI). IEEE Trans. Fuzzy Syst. (2017)
113. D. Yao, A method to standardize a reference of scalp EEG recordings to a point at infinity. Physiol. Meas. **22**, 693 (2001)
114. D. Yao, L. Wang, R. Oostenveld, K.D. Nielsen, L. Arendt-Nielsen, A.C. Chen, A comparative study of different references for EEG spectral mapping: the issue of the neutral reference and the use of the infinity reference. Physiol. Meas. **26**, 173 (2005)
115. H. Zeng, A. Song, R. Yan, H. Qin, EOG artifact correction from EEG recording using stationary subspace analysis and empirical mode decomposition. Sensors **13**, 14839–14859 (2013)
116. K. Zeng, D. Chen, G. Ouyang, L. Wang, X. Liu, X. Li, An EEMD-ICA approach to enhancing artifact rejection for noisy multivariate neural data. IEEE Trans. Neural Syst. Rehab. Eng. **24**, 630–638 (2016)
117. C. Zhang, J. Yang, Y. Lei, F. Ye, Single channel blind source separation by combining slope ensemble empirical mode decomposition and independent component analysis. J. Comput. Inf. Syst. **8**, 3117–3126 (2012)
118. C. Zhao, T. Qiu, An automatic ocular artifacts removal method based on wavelet-enhanced canonical correlation analysis, in *Abstracts of the Engineering in Medicine and Biology Society, EMBC, Annual International Conference of the IEEE* (2011)
119. Y. Zou, V. Nathan, R. Jafari, Automatic identification of artifact-related independent components for artifact removal in EEG recordings. IEEE J. Biomed. Health Inf. **20**, 73–81 (2016)

Chapter 3
EEG Spectral Analysis

Do-Won Kim and Chang-Hwan Im

Abstract Electroencephalogram (EEG) spectral analysis quantifies the amount of rhythmic (or oscillatory) activity of different frequency in EEGs. Based on numerous studies that reported significant relationship between the EEG spectrum and human behavior, cognitive state, or mental illnesses, EEG spectral analysis is now accepted as one of the principal analysis methods in the field of neuroscience. Despite the tremendous amount of research related to its usefulness, EEG spectral analysis still exhibits inconsistent results among studies. This might be partly because of the various methodological decisions the researchers have to make during EEG spectral analysis. Indeed, there is no standardized analysis procedure. In this chapter, we cover some background principles of spectral analysis and introduce important issues that researchers must consider during EEG spectral analysis.

3.1 Introduction

The human electroencephalogram (EEG) is one of the most complex sets of biomedical signals. It is believed to reflect a variety of processes of the brain, especially the neocortex, in which our cognitive function and sensorimotor information are processed. Since the first report of successful recording of the human EEG by neuropsychiatrist Hans Berger [6, 12], the EEG has been used extensively to help understand the functions of the brain. Although we still do not have enough understanding to describe every detail of the brain's dynamics, extensive neuroscience studies provide us with pieces of information on how the human brain operates in certain conditions, e.g., reaction to a given stimulus.

EEGs can be analyzed qualitatively, as Berger did in his report on the human EEG, or quantitatively, by using various computational EEG analysis methods. Qualitative

D.-W. Kim (✉)
Department of Biomedical Engineering, Chonnam National University, Yeosu, South Korea
e-mail: dowon.kim@jnu.ac.kr

C.-H. Im
Department of Biomedical Engineering, Hanyang University, Seoul, South Korea

C.-H. Im (ed.), *Computational EEG Analysis*, Biological and Medical Physics, Biomedical Engineering, https://doi.org/10.1007/978-981-13-0908-3_3

analysis is still used routinely in neurology and sleep studies, mainly to find abnormal signal patterns or roughly categorize well-known states of the brain that can be easily recognized by visual inspection [5, 18]. For instance, clinicians determine the origin of epileptic activity from ictal EEG recordings acquired during epileptic seizure [19, 38]. Another good example of qualitative EEG analysis is EEG-based sleep stage scoring, using visually evident patterns in the signal (e.g., slowing rhythm, existence of K-complex or spindle) [15, 37]. Naturally, the experience and expertise of the rater may influence the outcome, and this motivated researchers to develop less subjective measures. Quantitative analysis uses mathematical and statistical methods to find evident features to characterize the given EEG signal, even ones that are difficult to detect by visual inspection. Each approach classifies the EEG signal in terms of frequency or period, amplitude, phase relations, and morphology (waveform, topology, abundance, reactivity, and variability of these parameters) [7].

Among the various quantitative features of EEGs, one of the basic and common features is the frequency power of the EEG signal. From the earliest stages of EEG research, probably from the first report of the existence of alpha waves by Hans Berger, different velocities of the EEG recording were believed to reflect different states of the brain. Based on extensive research, now it is widely accepted that we can define distinct frequency bands with different roles and characteristics, namely, the delta (1–3 Hz), theta (4–7 Hz), alpha (8–12 Hz), beta (15–30 Hz), and gamma (30–100 Hz) frequency bands. The frequency ranges of each band slightly vary among different studies and researchers; however, it is generally accepted that a subtle difference (less than 1 Hz) in defining the frequency range does not make a significant difference [29]. Normally, the lower frequency band is believed to reflect subconscious states, mostly showing dominant activation during deep sleep or drowsiness, while relatively higher frequency bands reflect more alerted, active states or are associated with higher cognitive functions [20, 35].

Two important factors are mainly considered in traditional EEG spectral analysis: the amount (usually reported as *power*) of a specific frequency band and its spatial distribution—see Fig. 3.1a. The *power* represents the amount of the frequency band included in the signal, where both increase and decrease of EEG power are meaningful information to understand the underlying brain function. The spatial distribution of the power is also considered to be crucial, because the power changes in different brain areas may represent different processes of the brain. EEG spectral powers are generally represented as topological distributions on the scalp surface, and this is usually referred to as *quantitative EEG* (qEEG) analysis, allowing for intuitive comparison among groups or conditions in clinical applications. For instance, traditional qEEG analysis was done by comparing the individual data with a normative database [22, 23]. The normative database basically contains frequency band power data carefully collected by several hundreds of healthy subjects with diverse ages. The spectral power of the individual EEG is usually converted to z-score, which highlights whether the spectral powers of any EEG channel are enhanced or reduced compared with the normative database.

In the initial stage of development of spectral analysis methods, it was not easy to track the temporal variations of the frequency spectrum, because the *temporo-*

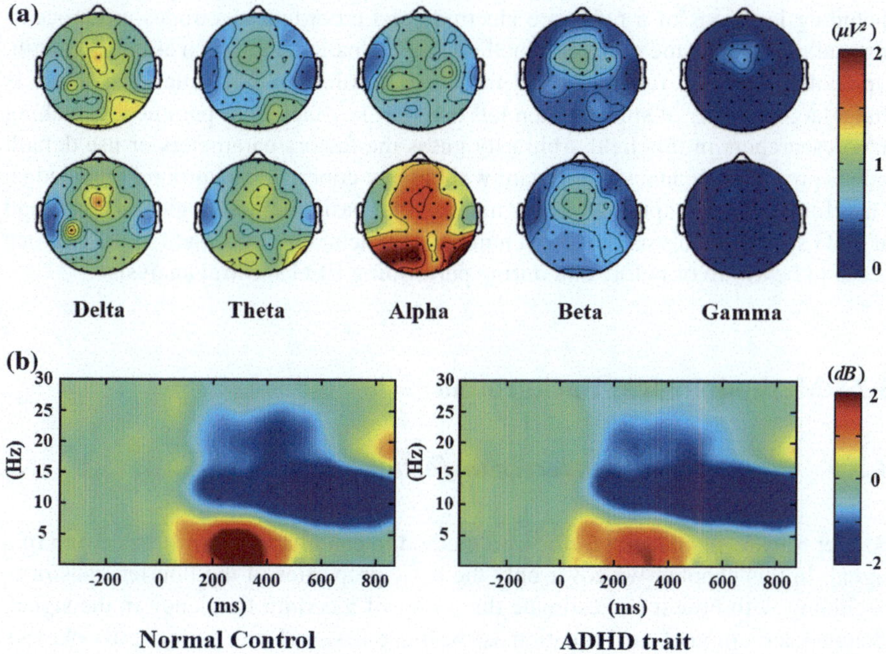

Fig. 3.1 **a** Examples of topographies showing spatial distributions of average absolute spectral powers for five frequency bands: delta (1–5 Hz), theta (5–8 Hz), alpha (8–12 Hz), beta (15–30 Hz), and gamma (30–55 Hz). The EEGs were recorded from a normal person in resting state with eyes open (top row) and eyes closed (bottom row). **b** Examples of time–frequency maps: grand average of time–frequency spectra from 20 normal controls (left) and 20 adults with attention deficit/hyperactivity disorder (ADHD) (right)

spectral EEG analyses need sufficient length of data to secure minimum frequency resolution. Therefore, early spectral analysis studies were mostly done on long-lasting and stable EEG recordings, e.g., EEG acquired during resting state or sleep, for which the exact timing is relatively less important compared with time-locked or stimulus-dependent experiments. However, after the mathematical background for estimating the spectral power in a short time window was established [42], it became possible to investigate the temporal changes of EEG power over a relatively short time interval, allowing for investigations of time–frequency dynamics of EEGs with respect to external or internal cues—see Fig. 3.1b.

Spectral analysis is a fundamental computational EEG analysis method that can provide information on power, spatial distribution, or event-related temporal change of a frequency of interest. However, EEG spectral analysis often has been regarded as an unreliable and imprecise method by some neuroscientists and clinicians [29] owing to the inconsistent results among studies that used spectral analysis. The inconsistency is partly due to the absence of a *golden standard* in the analysis procedure [22]. The researchers confront a series of choices of experimental factors, including locations of a reference electrode and recording electrodes, preprocessing methods, and some necessary methodological parameters, such as epoch length, types of windowing functions, and frequency resolution. What makes it worse is that a large number of studies often fail to report key factors or parameters, making new researchers in this field arbitrarily guess the factors/parameters or use default values provided by analysis software without any concrete background knowledge. Therefore, in this chapter, we cover not only the basic methodological background of EEG spectral analysis but also a number of crucial concepts and factors that one needs to be aware of before and during performing EEG spectral analysis.

3.2 Methodological Background

3.2.1 Continuous Fourier Transform

Fourier transform is a straightforward method to calculate the power spectrum of a signal. In this chapter, we cover only the basic properties of the Fourier transform, beginning with how it can estimate the power of a certain frequency in the signal. Detailed derivation of each equation can be found in signal-processing textbooks [30, 34]. First, we start with the general case of a continuous nonperiodic signal. Assume $x(t)$ as a signal that is infinite in length and continuous in time. The *continuous Fourier transform* (CFT) of a function $x(t)$ is then defined as

$$X(\omega) = \int_{-\infty}^{+\infty} x(t)e^{-j\omega t}dt, \tag{3.1}$$

where $e^{-j\omega t}(= \cos \omega t - j \sin \omega t)$ are the complex exponentials, and ω is the angular frequency corresponding to the linear frequency f ($\omega = 2\pi f$). Equation (3.1) quantifies the amount of contribution of each frequency ω in constituting the original signal. If the signal $x(t)$ is defined for all real numbers t, for any $\omega \in \mathbf{R}$, integrating $x(t)$ against $e^{-j\omega t}$ with respect to t produces a complex-valued function of ω. The square magnitude of $X(\omega)$ ($|X(\omega)|^2$) is called the *power spectrum* or *power spectral density* (PSD), where the angle $\angle X(\omega)$ denotes the phase at the given frequency ω. In the frequency domain analysis of EEG, our major interest is the power spectrum of a given EEG signal, which contains information on how much each frequency component is contained in the given signal.

The definition in (3.1) can also be considered as a correlation between the signal $x(t)$ and the complex sinusoidal functions $e^{-j\omega t}$. Therefore, Fourier transform can be more intuitively understood as representing how similar the given signal is to the complex exponential of a given frequency. The higher the correlation is with the sinusoidal with frequency ω, the more influence of the frequency ω exists in the original signal $x(t)$.

Using the frequency domain characteristic of $x(t)$, we can also transform the signal spectrum back to the time series signal using the *inverse Fourier transform*. The inverse Fourier transform is defined as

$$x(t) = \frac{1}{2\pi} \int_{-\infty}^{+\infty} X(\omega) e^{j\omega t} d\omega. \tag{3.2}$$

3.2.2 Discrete Fourier Transform

The CFT described in the previous section assumes that the signal is continuous in time and infinite in length. However, in any modern EEG recording, the signal is recorded in a limited time interval and stored digitally after it is amplified. In other words, the recorded EEG signal that we want to analyze is neither infinite nor continuous in time. Therefore, we use the *discrete Fourier transform* (DFT) instead of the CFT. The DFT assumes that its input signal is one period of a periodic signal. Its output is the discrete frequency spectrum of this periodic signal.

Consider a discrete signal $x[n]$, the length of which is finite and $n = 1, 2, \ldots, N$. The signal is derived by sampling a continuous signal $x(t)$ with an equal time interval Δt or a sampling frequency $f_s = 1/\Delta t$. The signal length is fixed to a finite length of $T = N\,\Delta t$. The discrete form of the Fourier transform is defined as:

$$X[k] = \sum_{n=0}^{N-1} x[n] e^{-j2\pi kn/N}, \text{ where } k = 0, \ldots, N-1. \tag{3.3}$$

The DFT gives the frequency spectrum at discrete frequencies f_k, where the following relationship is given:

$$f_k = \frac{k}{N\Delta t}, \tag{3.4}$$

when the frequency resolution of the spectral density is given by

$$\Delta f = \frac{1}{T} = \frac{1}{N\Delta t}. \tag{3.5}$$

The frequency resolution is determined only by the length (number of samples) of the signal. According to the Shannon sampling theorem [36], the highest frequency of the power spectrum is limited to the Nyquist frequency f_N given by

$$f_N = \frac{f_s}{2} = \frac{1}{2\Delta t}. \tag{3.6}$$

It is important to understand that the sampling frequency f_s is the only factor that determines the Nyquist frequency.

As in the CFT, the Fourier coefficients $X[k]$ are complex numbers. Therefore, $X[k]$ can be expressed as the following in Cartesian form:

$$X[k] = X_{\mathrm{Re}}[k] + jX_{\mathrm{Im}}[k], \tag{3.7}$$

where $X_{\mathrm{Re}}[k]$ and $X_{\mathrm{Im}}[k]$ are the real and imaginary parts of $X[k]$, respectively. From (3.7), the magnitude $|X[k]|$ and the phase $\Phi[k]$ can be expressed as

$$|X[k]| = \sqrt{X_{\mathrm{Re}}^2[k] + X_{\mathrm{Im}}^2[k]}, \tag{3.8}$$

$$\Phi[k] = \tan^{-1}\frac{X_{\mathrm{Im}}[k]}{X_{\mathrm{Re}}[k]}. \tag{3.9}$$

The inverse DFT transforms the Fourier coefficient $X[k]$ back into the discrete time series $x[n]$, as follows:

$$x[n] = \frac{1}{N}\sum_{n=0}^{N-1} X[k]e^{j2\pi kn/N}. \tag{3.10}$$

3.2.3 Fast Fourier Transform

The *fast Fourier transform* (FFT) is a particular implementation of the DFT that gives identical results with reduced calculations [10]. The calculation of (3.4) requires N^2 complex multiplications, because, for each of the N discrete frequencies, it is necessary to calculate the sum of N multiplications of complex exponentials. However, in cases when N is a power of 2 (e.g., 64, 128, 512, 1024, …), many of these multiplications result in identical values, and many of the complex exponentials are zero or 1. When redundant computations are removed, the number of multiplications is reduced to $N\log_2(N)$ rather than N^2, reducing the computational burden, especially when N is large. Spectral estimation based on the FFT assumes that the signal is stationary and slowly varying.

3.2.4 Aliasing and Leakage

Spectral estimation based on the FFT has intrinsic properties called *aliasing* and *leakage* [28], which need to be considered carefully. To understand the aliasing, consider a continuous signal with a single frequency, as shown in Fig. 3.2. As the signal is recorded, the original signal is sampled to a discrete-time signal, depending on the sampling frequency (or interval) of the analog-to-digital converter (ADC). If the

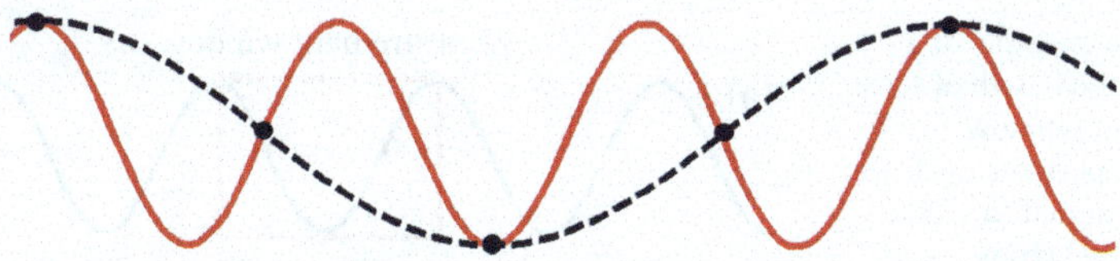

Fig. 3.2 An example of an aliasing effect. If the original signal (red bold) is sampled with a sampling frequency that does not satisfy the Shannon Sampling Theorem, the sampled signal (black dotted) will not able to reconstruct the original signal, but rather it will be presented as a signal with a spurious frequency

signal is sampled with a relatively high sampling frequency that fulfills the Shannon sampling theorem, the frequency domain characteristics of the original signal can be fully reconstructed. However, if the signal is sampled at a sampling frequency lower than the Nyquist frequency of the original signal (e.g., five black dots in Fig. 3.2), the frequency information of the original signal cannot be fully reconstructed or estimated solely using the sampled signal. This *undersampling* rather creates an activation in a different frequency (see the dotted line in Fig. 3.2), thereby resulting in a phantom or spurious power at a frequency that is not present in the original signal. This effect is called *aliasing*. To avoid the aliasing effect, antialiasing low-pass filters should be applied before the digitization of the signal. Antialiasing filters restrict the bandwidth of the original signal so that the sampling frequency of the system can fulfill the Shannon sampling theorem.

Another important issue to be considered in the evaluation of the power spectral density of a finite signal using FFT is the existence of *spectral leakage*. As mentioned before, the DFT assumes that the input signal is one period of a periodic signal. Consider a periodic signal, as shown in Fig. 3.3a, and assume that we have recorded a *window* of the given signal, as shown in Fig. 3.3b. If the window length is the same as the periodic cycle of the original signal, as in Fig. 3.3c, the power spectra of the original and repeated signals are identical. However, if the repeated signal, as in Fig. 3.4c, of the truncated signal, shown in Fig. 3.4b, has some discontinuities in the time domain, the frequency spectrum gets attenuated, i.e., the original power spectrum spreads out to nearby frequencies.

Therefore, the simplest way to avoid the spectral leakage would be to decide carefully the length of the measuring window, so the repeated signal does not have any discontinuity. This is possible only if we know the exact frequency composition of the recording signal or if the signal is periodic, which is usually impossible in real recording situations where multiple frequency components are mixed together.

For example, Fig. 3.5a is a 5 Hz sine wave of a 2 s window including 10 full cycles of sinusoids. The power spectrum of the signal is presented in Fig. 3.5b, which correctly shows the spectrum of the original signal. However, a signal in Fig. 3.5c has identical amplitude, phase, and frequency characteristics to those of the signal in Fig. 3.5a, but it includes a noninteger number of cycles (reduced window size), leading to a discontinuous waveform when it is repeated. This, consequently, results in leakage in the power spectrum when Fourier-based methods are used—see

Fig. 3.3 An example of a case where the length of the measurement window matches the periodic cycle of the original signal (**a**). The captured signal (**b**) is assumed to repeat itself in time and creates the identical continuous signal (**c**)

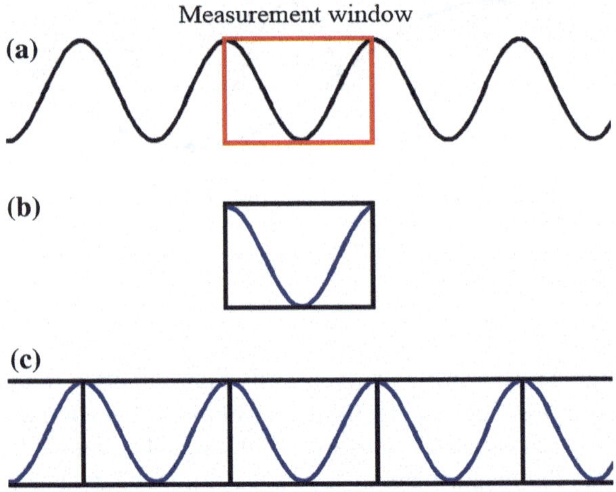

Fig. 3.5d. The power is spread out around the original frequency (5 Hz), and the sum of the power does not match the total power of the original signal. The leakage effect can be reduced to some extent by using an appropriate windowing function. Figure 3.5e shows a signal after multiplying the signal in Fig. 3.5c by a *Hanning window* function with the same length. Figure 3.5f shows the frequency spectrum of the signal in Fig. 3.5e, where it can be observed that the leakage shown in Fig. 3.5d was somewhat reduced. A more detailed description of the windowing functions can be found in Sect. 3.3.3.

Fig. 3.4 An example of a case where the length of the measurement window does not match the periodic cycle of the original signal (**a**). The captured signal (**b**) is assumed to repeat itself in time but creates a discontinuous signal (**c**) that is not identical to the original signal (**a**)

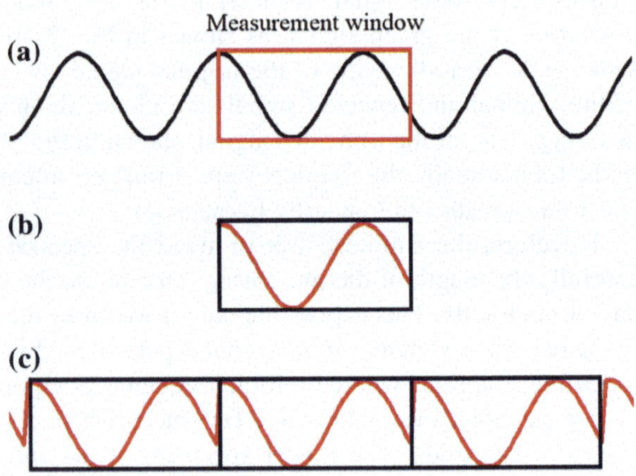

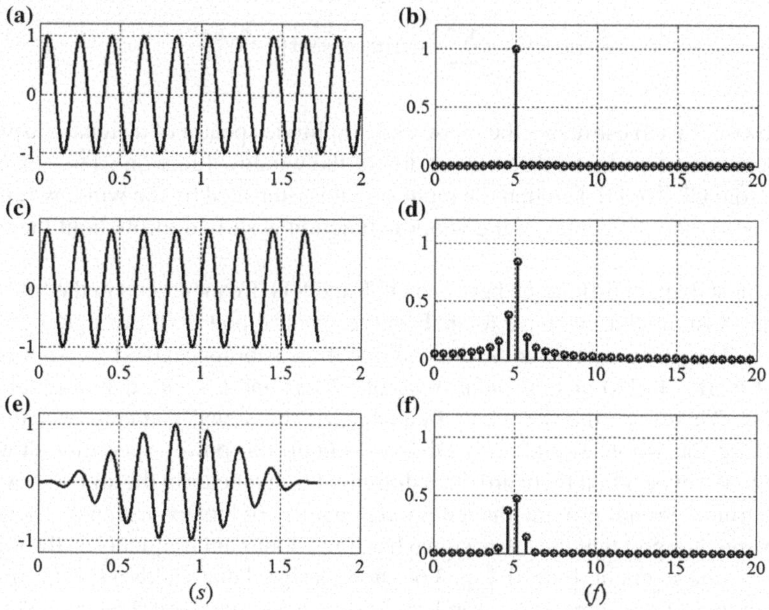

Fig. 3.5 **a** A 5–Hz sinusoidal signal over 2 s and **b** its power spectrum calculated using Fourier transform. **c, d** A sinusoidal signal with noninteger cycles can cause power leakage around the main peak. **e, f** The leakage can be reduced using proper windowing functions, such as the Hanning window

3.2.5 Short-Time Fourier Transform

DFT calculates the spectral composition of a finite time interval. It provides a good estimate of the frequency spectrum of the given signal; however, it is not easy to track the temporal change of the power spectrum over a relatively short time period. Since EEG is generally considered a nonstationary and time-variant signal, a non-negligible amount of temporal details is lost by the spectral analysis based on the DFT. To address this issue, Welch [42] proposed a method called *short-time Fourier transform* (STFT), the mathematical definition of which is given as

$$X(\tau, \omega) = \int_{-\infty}^{+\infty} x(t)w(t - \tau)e^{-j\omega t}dt. \tag{3.11}$$

STFT is identical to the CFT of a signal $x(t)w(t - \tau)$, where $w(t - \tau)$ is a window function that is shifted by τ. Now, we have a two-dimensional output $X(\tau, \omega)$ that estimates the spectra for both time and frequency, where the window is centered at $t = \tau$. The discrete version of the STFT can be written as

$$X[m,k] = \sum_{n=-\infty}^{+\infty} x[n]w[n-m]e^{-j2\pi kn/N}. \quad (3.12)$$

Likewise, (3.12) estimates the phase and amplitude spectra of a signal $x[n]w[n-m]$, where $w[n-m]$ is the discrete version of the window function $w(t-\tau)$. STFT differs from CFT or DFT in that the input signal is truncated by the window function $w(t)$ or $w[n]$, and the center of the window function is shifted throughout the whole signal.

The most distinct difference between DFT and STFT can be seen in the following example. Assume that we have to analyze the spectral power of the signal given in Fig. 3.6a. The given signal can be divided into three subranges: (0–1 s), a sinusoidal wave of 6 Hz; (1–2 s), a sinusoidal wave of 17 Hz; and (2–3 s), a sinusoidal wave of 31 Hz. Therefore, the dominant frequency of the signal changes over time. If we analyze the whole signal with DFT, we obtain the power spectrum shown in Fig. 3.6b. We expect that there are three dominant frequencies in the signal; however, the spectrum does not provide the temporal dynamics of the given signal. Therefore, we may misinterpret that the given signal has three dominant frequencies, throughout the entire time segment. Indeed, Fig. 3.6c shows a signal that yields a similar spectral profile to that of the signal shown in Fig. 3.6a. Strictly speaking, however, the DFT of Fig. 3.6c will have fewer sidelobes than that in Fig. 3.6a; they are mainly caused by the discontinuity of the signal at $t = 1$ and $t = 2$. Figure 3.6d shows a result of the STFT analysis of the same signal shown in Fig. 3.6a. The x- and y-axis denote time and frequency, respectively, and the power is color-coded, following (3.12). We can observe the temporal dynamics of the given signal—the dominant frequency of the signal changes every 1 s.

3.3 Practical Remarks on EEG Spectral Analysis

3.3.1 Choosing Adequate Sampling Frequency

According to the Shannon sampling theorem, the sampling frequency must be at least twice the frequency that needs to be observed. For instance, to precisely estimate the spectral power of the gamma band (30–60 Hz), the sampling frequency must be at least 120 Hz. Researchers should keep in mind that the Shannon sampling theorem provides the minimum restriction for choosing the sampling frequency. In practice, it is generally recommended to set a sampling frequency higher than the Nyquist frequency, mostly three times larger than the maximum frequency of interest [2], considering the transition band of the antialiasing low-pass filter.

Modern EEG amplifiers usually provide options to adjust the sampling frequency based on user demand, mostly from 256 to 2048 Hz. The sampling frequency does not influence the frequency resolution of the spectrum, but it has a direct relationship with the amount of data storage needed for each recording. The higher the sampling

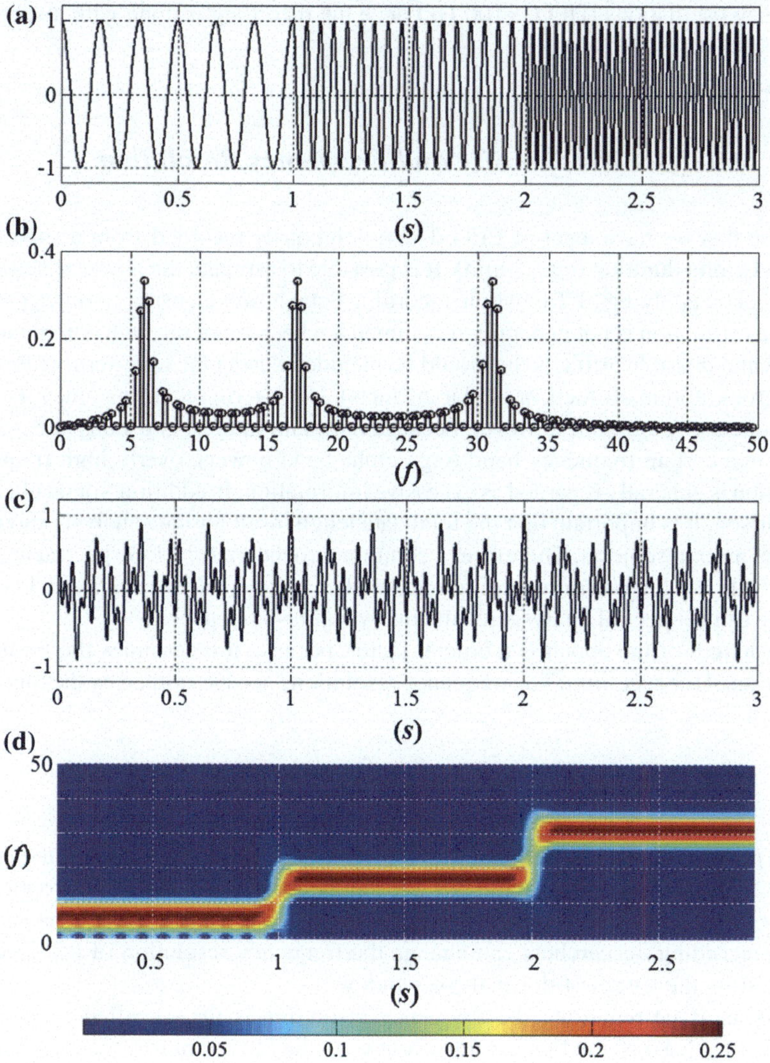

Fig. 3.6 **a** A sample signal with different dominant frequencies over time; 6 Hz for 0–1 s, 17 Hz for 1–2 s, and 31 Hz for 2–3 s. **b** Power spectrum evaluated using the whole signal of (**a**). **c** A signal with 6, 17, and 31 Hz components simultaneously, which has a power spectrum like that in (**b**). **d** STFT result of signal (**a**), where the temporal dynamics of the signal are accurately represented

frequency, the more data storage is required. Therefore, long-term EEG recordings (e.g., sleep EEG or epileptic EEG monitoring) tend to use relatively lower sampling frequencies compared with short-term recordings, such as resting-state EEG [3]. Keep in mind that *downsampling* is always possible, but *upsampling* is impossible once the signal recording is done. For instance, it is impossible to observe high frequency oscillations (HFO, >100 Hz) once the recording is done with a sampling rate of 200 Hz.

3.3.2 Analysis Window Size and Frequency Resolution

Suppose that we have a set of EEG data continuously recorded from a subject for relatively long duration (e.g., 2 min). It is possible to estimate the power spectrum of the signal by applying FFT to whole recording data; however, using so much data for spectral estimation is not generally recommended because of the high computational burden and possible artifacts that could be included during the long-term recordings (e.g., gross motion artifacts or ocular artifacts). Long-term recordings may increase the frequency resolution; however, because our main interest is usually the average power in a certain frequency band (e.g., alpha band power), overly high frequency resolution is generally regarded as excessive information. In addition, for an objective comparison, it is important that the total data length used for data analysis should be uniform among subjects. Therefore, a common procedure for EEG spectral analysis is to divide the long-term recording into smaller pieces, called *epochs*, and take an average of the spectral analysis results over *artifact-free* epochs [32].

The length of the epoch is a crucial factor, because it determines the frequency resolution of the spectrum. The frequency resolution f_c is determined by the following equation:

$$f_c = \frac{f_s}{N}, \tag{3.13}$$

where f_s is the sampling frequency, and N is the number of samples in the epoch. Hence, for a fixed sampling frequency, the number of samples, which is proportional to time, determines the frequency resolution. Because the sampling frequency is fixed before recording, researchers can change the frequency resolution of the spectrum by adjusting the length of the analysis window.

Another issue regarding the frequency resolution is the so-called *picket fence effect* [21, 31]. Since the Fourier transform is applied to a sampled (discrete) signal, the spectrum has values only at discrete frequency samples that are multiples of the fundamental frequency f_c. Therefore, the spectral information is accurate only when the frequency of the original signal matches nf_c ($n = 0, 1, 2, \ldots, N-1$); otherwise, *leaked* frequency components can be observed in adjacent frequency samples. For instance, suppose a signal has a convex spectrum peaking at f_1, as shown in Fig. 3.7a. With the frequency resolution f_c, the main peak of the signal power spectrum can be correctly represented at f_1. However, if the peak of the spectrum is located at f_2, which is the middle point of f_1 and the next frequency sample (=($f_1 + f_c$)/2), the spectral characteristics of the signal cannot be correctly estimated, as shown in Fig. 3.7b.

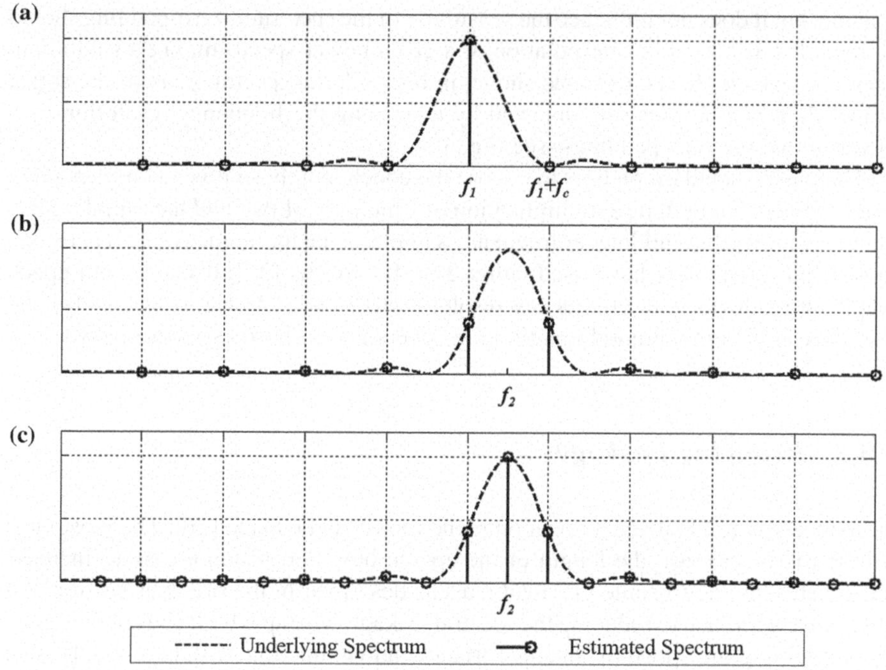

Fig. 3.7 **a** An illustration of a discrete power spectrum (solid stem) when the frequency resolution can correctly estimate the main peak of the underlying spectrum (dotted line). **b** However, if the frequency resolution is not enough to estimate the spectrum, leakage might occur to nearby frequency bins. **c** This picket-fence effect can be circumvented by increasing frequency resolution using zero padding

This is a common problem arising when one wants to use a window size of 2^N for analyzing EEG data recorded with an EEG system that does not provide sampling frequencies of the form of 2^N (250, 500, 1000 Hz, etc.). For example, assume that we want to evaluate the alpha band (8–12 Hz) power of an EEG signal recorded at a sampling frequency of 250 Hz. If the analysis window size is set to be 512 samples (=2.048 s) to take the full advantage of FFT, the resultant power spectrum would have discrete values at every multiple of 0.448 Hz (=250/512). Now, it is difficult to decide how to evaluate the average alpha band power, because the spectrum does not provide power values at 8 and 12 Hz. The available frequency samples adjacent to 8 Hz are 7.813 and 8.301 Hz, whereas those adjacent to 12 Hz are 11.719 and 12.207 Hz. Therefore, we must make a choice among possible frequency pairs; the choice might result in some differences in the band power estimates.

Zero-padding before FFT can be a possible solution to address the issues presented in the previous paragraph [31]. Zero-padding is a simple concept; it refers to adding a series of zeros to the end of an epoch to increase the length of the epoch. Zero-padding can increase the frequency resolution, thereby matching the required samples for FFT. However, it should be noted that zero-padding decreases only the frequency spacing, but it does not influence the sensitivity of the spectrum; zero-padding should be regarded as a kind of interpolation of a given power spectrum, not a solution to decrease leakage. In the example shown in Fig. 3.7, the spectral peak of the signal in Fig. 3.7b can be precisely detected by increasing the frequency resolution by a factor of two via zero-padding—see Fig. 3.7c.

There is no standard on how to choose the epoch length, so it is recommended to make the epoch length be a multiplication of a multiple of two and the sampling frequency. This enables all integer frequency samples to be harmonics of f_c. Therefore, most EEG researchers have used either 2 or 4 s epochs [25], because both epoch lengths provide adequate frequency resolution (0.5 and 0.25 Hz, respectively) and also have sufficient temporal margin to reject epochs with excessive artifacts.

3.3.3 Reducing Leakage

To reduce spectral leakage, two recommendations can be made [26]. The most ideal solution is to increase the length of the epoch; however, it is not easy to increase the epoch size due to some practical reasons described in the previous section. An alternative approach to reducing the spectral leakage is to taper both ends of the epoch by multiplying the epoch by an appropriate window function [40]. Figure 3.4e is an example of applying a Hanning window on the same signal shown in Fig. 3.4c. As shown in Fig. 3.4f, the use of a Hanning window could effectively reduce the spectral leakage. This procedure is known as *windowing*. There are many types of windowing functions, such as rectangular, Barlett, Hanning, Hamming, and Backman [13]; each window has its own characteristic to shape the spectrum—Table 3.1; see Prabhu et al. [33] for the detailed performance comparison among window functions.

It should be noted that the window function must be used before the signal is padded with zeros, because window functions are used to smooth the endpoints of the truncated signal while preserving the spectral power of the original signal.

3.3.4 Window Function for STFT

Two factors must be carefully determined for STFT analysis: the type and size of the window function. Various types of window function can be considered, such as rectangular, Hanning, Hamming, Gaussian, and Blackman, all of which can be used to suppress spectral leakage but are slightly different from each other in their performance [33]. Among them, Hanning and Hamming windows have been most widely used in EEG spectral analysis. However, the length of the window function is more difficult to decide. As the window size increases, the frequency resolution also increases, but the analysis results become less sensitive to time. Use of a short window size would provide good temporal resolution but give poor frequency resolution. The window should be narrow enough to ensure that the signal truncated by the window is

Table 3.1 Common windowing functions and their characteristics, adapted from Prabhu et al. [33]

Name	Function	Peak side-lobe amplitude (dB)	Mainlobe width	Minimum Stopband Attenuation (dB)
Rectangular	$\omega(n) = 1,\ 0 \leq n \leq N-1$	−13	$4\pi/N$	−21
Barlett	$\omega(n)\begin{cases} 2/N,\ 0 \leq n \leq (N-1)/2 \\ 22n/N,\ (N-1)/2 \leq n \leq N-1 \end{cases}$	−25	$8\pi/N$	−25
Hanning	$\omega(n) = 0.5 \times (1 - \cos(2\pi n/N),\ 0 \leq n \leq N-1$	−31	$8\pi/N$	−44
Hamming	$\omega(n) = 0.54 - 0.46\cos(2\pi n/N),\ 0 \leq n \leq N-1$	−43	$8\pi/N$	−53
Backman	$\omega(n) = 0.42 - 0.5\cos(2\pi n/N) + 0.08\cos(4\pi n/N),\ 0 \leq n \leq N-1$	−57	$12\pi/N$	−74

stationary, but it should not be too narrow to secure adequate frequency resolution for analysis. If the window size does not fulfill the desired frequency resolution, consider using zero-padding to increase the frequency resolution, but not excessively, because the result can be distorted by chance. The temporal resolution of STFT can also be increased by shifting the window in a small step over time. The amount of shift of the window over time can also be defined as the *overlap ratio* of the window. As the overlap ratio increases, fewer samples of the upcoming signal are included in the window. Also, the temporal resolution will increase as the overlap ratio increases; therefore, the frequency change over time shows a *continuous-like* function; however, using too much the overlap will increase computational burden.

3.3.5 Absolute Power Versus Relative Power

Absolute power is a measure directly indicating the amount of spectral power of a specific frequency, and it is straightforward to interpret. However, the spectral pattern and/or the overall power of EEG are diverse, not only among age groups, gender, or cognitive states (drowsiness, sleep, attention, etc.) [9, 14, 41], but also among individuals under the same experimental conditions [17, 27]. Also, the power spectrum of the EEG shows an exponential decrease with increasing frequency, which means the lower-frequency components, such as the delta and theta band activities, are much bigger than the higher-frequency components, such as alpha and beta band activities [4]; therefore, the absolute power might not effectively detect the small changes in higher frequencies. Moreover, it is generally difficult to compare datasets recorded with different EEG amplifiers because of the unique frequency response of each amplifier. Therefore, it is sometimes useful to evaluate the *relative power* of each frequency band by dividing the absolute power of each band by the total power (usually, 1–50/60 Hz) or by the sum of powers in the frequency bands of interest. The relative power represents the proportion of each band to the given signal. According to Klimesch [24], the relative power of the alpha band highlighted the increase of alpha power in children with respect to age, although it was hard to observe such a trend in absolute power.

In comparison with absolute power, relative power could be an intuitive indicator to track changes in overall dominance of each band over time or under different conditions. However, some researchers insist that relative power may lead to misinterpretation of the data. Indeed, an increase of relative power in one frequency band could be interpreted as a decrease of relative power in another band, even when the absolute power of the latter frequency band did not change [8]. In addition, the definition of frequency bands is not standardized, making it difficult to compare the analysis results among studies. Nevertheless, both absolute and relative powers contributed to many important findings in the field of neuroscience; therefore, they should be treated equally.

3.3.6 Other Considerations in Spectral Analysis

There is no restriction in the number of electrodes used for spectral analysis; however, at least 19 (according to the international 10–20 system) or more electrodes are recommended to observe the overall spatial distribution of spectral power for each frequency band [2]. It is common to use reference electrodes placed at electrically neutral places, such as nose tip reference and linked-ear reference; however, some researchers prefer to use reference-free methods, such as common average reference (CAR). Physical reference electrodes often suffer from local contamination but have an advantage that each electrode can be treated independently. On the other hand, CAR generally requires a number of evenly distributed electrodes on the scalp surface to fulfill its mathematical assumption [11]. Moreover, CAR is sensitive to an artifact (e.g., eye blink artifact and electromyogram artifact), because the artifact is evenly distributed to other electrodes. Therefore, if one decides to use CAR, it is important to reject epochs including artifacts or remove/reduce the artifacts before the application of CAR using signal processing procedures, e.g., independent component analysis (ICA).

In practical EEG analysis, it is a widely accepted process to group a few adjacent electrodes within a specific region of interest and calculate the average spectral power of the region. Division of the regions is generally made according to hemispheres (left/right) or lobes (frontal/central/temporal/occipital) or by combining both. Grouping the electrodes by regions or hemispheres can better highlight regional differences—for instance, alpha asymmetry [16] of left and right hemispheres. Averaging the spectral power in small areas might be reasonable, considering the low spatial resolution of EEGs. Moreover, grouping nearby electrodes is advantageous not only to increase the reliability of the power spectrum by averaging, but also to relieve the necessity of multiple-testing correction by decreasing the number of simultaneous comparisons. Likewise, the basic frequency bands (delta, theta, alpha, beta, and gamma) can also be divided into multiple subbands. Alpha and beta bands are sometimes fragmented by (1) the low-alpha and high-alpha and (2) the low-beta and high-beta subbands, to investigate the functions of more specific bands [1, 39].

References

1. L.I. Aftanas, S.A. Golocheikine, Human anterior and frontal midline theta and lower alpha reflect emotionally positive state and internalized attention: high-resolution EEG investigation of meditation. Neurosci. Lett. **310**, 57–60 (2001). https://doi.org/10.1016/S0304-3940(01)02094-8
2. American Clinical Neurophysiology, Guideline fourteen: guidelines for recording clinical EEG on digital media. J. Clin. Neurophysiol. **11**, 114–115 (1994)
3. American Clinical Neurophysiology, Guideline twelve: guidelines for long-term monitoring for epilepsy. J. Clin. Neurophysiol. **11**, 88–110 (1994)
4. J.S. Barlow, *The Electroencephalogram: Its Patterns and Origins* (MIT Press, Cambridge, 1993), pp. 213–220
5. R.J. Barry, A.R. Clarke, S.J. Johnstone, A review of electrophysiology in attention-deficit/hyperactivity disorder: I. Qual. Quant. Electroencephalogr. Clin. Neurophysiol. **114**, 171–183 (2003). https://doi.org/10.1016/S1388-2457(02)00362-0
6. H. Berger, Über das Elektrenkephalogramm des Menschen. Archiv für Psychiatrie und Nervenkrankheiten **87**, 527–570 (1929). https://doi.org/10.1007/bf01797193
7. M.A. Brazier, Preliminary proposal for an EEG terminology by the Terminology Committee of the International Federation for Electroencephalography and Clinical Neurophysiology. Electroencephalogr. Clin. Neurophysiol. **13**, 646–650 (1961)
8. T.H. Budzynski, H.K. Budzynski, J.R. Evans, A. Abarbanel, *Introduction to Quantitative EEG and Neurofeedback: Advanced Theory and Applications* (Elsevier, 2009)
9. J. Carrier, S. Land, D.J. Buysse, D.J. Kupfer, T.H. Monk, The effects of age and gender on sleep EEG power spectral density in the middle years of life (ages 20–60 years old). Psychophysiology **38**, 232–242 (2001). https://doi.org/10.1111/1469-8986.3820232
10. J.W. Cooley, J.W. Tukey, An algorithm for the machine calculation of complex Fourier series. Math. Comput. **19**, 297–301 (1965). https://doi.org/10.2307/2003354
11. J. Dien, Issues in the application of the average reference: review, critiques, and recommendations. Behav. Res. Methods Instrum. Comput. **30**, 34–43 (1998). https://doi.org/10.3758/bf03209414
12. G. Dietsch, Fourier-analyse von Elektrencephalogrammen des Menschen Pflüger's Archiv für die gesamte. Physiologie des Menschen und der Tiere **230**, 106–112 (1932). https://doi.org/10.1007/bf01751972
13. P. Djuric, S. Kay, Spectrum Estimation and Modeling, in *The Digital Signal Processing Hanbook*, 2nd edn., ed. by V. Madisetti (CRC Press, Boca Raton, 2017)
14. R.E. Dustman, D.E. Shearer, R.Y. Emmerson, Life-span changes in EEG spectral amplitude, amplitude variability and mean frequency. Clin. Neurophysiol. **110**, 1399–1409 (1999). https://doi.org/10.1016/S1388-2457(99)00102-9
15. I. Feinberg, R.L. Koresko, N. Heller, EEG sleep patterns as a function of normal and pathological aging in man. J. Psychiatr. Res. **5**, 107–144 (1967). https://doi.org/10.1016/0022-3956(67)90027-1
16. I.H. Gotlib, EEG alpha asymmetry, depression, and cognitive functioning. Cogn. Emot. **12**, 449–478 (1998). https://doi.org/10.1080/026999398379673

17. S. Gudmundsson, T.P. Runarsson, S. Sigurdsson, G. Eiriksdottir, K. Johnsen, Reliability of quantitative EEG features. Clin. Neurophysiol. **118**, 2162–2171 (2007). https://doi.org/10.1016/j.clinph.2007.06.018
18. T.M. Itil, Qualitative and quantitative EEG findings in schizophrenia. Schizophr. Bull. **3**, 61–79 (1977). https://doi.org/10.1093/schbul/3.1.61
19. H. Jaseja, B. Jaseja, EEG spike versus EEG sharp wave: differential clinical significance in epilepsy. Epilepsy Behav. **25**, 137 (2012). https://doi.org/10.1016/j.yebeh.2012.05.023
20. O. Jensen, J. Kaiser, J.-P. Lachaux, Human gamma-frequency oscillations associated with attention and memory. Trends Neurosci. **30**, 317–324 (2007). https://doi.org/10.1016/j.tins.2007.05.001
21. B.W. Jervis, M. Coelho, G.W. Morgan, Spectral analysis of EEG responses. Med. Biol. Eng. Compu. **27**, 230 (1989). https://doi.org/10.1007/bf02441479
22. D.A. Kaiser, QEEG: State of the art, or state of confusion. J. Neurother. **4**, 57–75 (2000). https://doi.org/10.1300/J184v04n02_07
23. D.A. Kaiser, Basic principles of quantitative EEG. J. Adult Dev. **12**, 99–104 (2005). https://doi.org/10.1007/s10804-005-7025-9
24. W. Klimesch, EEG alpha and theta oscillations reflect cognitive and memory performance: a review and analysis. Brain Res. Rev. **29**, 169–195 (1999). https://doi.org/10.1016/S0165-0173(98)00056-3
25. M.J. Levy, Effect of epoch length on power spectrum analysis of the EEG. Anesthesiology **66**, 489–495 (1987)
26. V. Madisetti, D.B. Williams, *The Digital Signal Processing Handbook. The Electrical Engineering Handbook Series* (CRC Press; IEEE Press, Boca Raton; New York, 1998)
27. S. Makeig, M. Inlow, Lapse in alertness: coherence of fluctuations in performance and EEG spectrum. Electroencephalogr. Clin. Neurophysiol. **86**, 23–35 (1993). https://doi.org/10.1016/0013-4694(93)90064-3
28. J. Muthuswamy, N.V. Thakor, Spectral analysis methods for neurological signals. J. Neurosci. Methods **83**, 1–14 (1998). https://doi.org/10.1016/S0165-0270(98)00065-X
29. M.R. Nuwer, D. Lehmann, F.L. da Silva, S. Matsuoka, W. Sutherling, J.F. Vibert, IFCN guidelines for topographic and frequency analysis of EEGs and EPs. The International Federation of Clinical Neurophysiology. Electroencephalogr. Clin. Neurophysiol. Suppl. **52**, 15–20 (1999)
30. A. Oppenheim, R. Schafer, *Discrete-Time Signal Processing*, 3rd ed. (Prentice-Hall Signal Processing Series) (Prentice Hall, 2009). Doi: citeulike-article-id:10412772
31. J. Pardey, S. Roberts, L. Tarassenko, A review of parametric modelling techniques for EEG analysis. Med. Eng. Phys. **18**, 2–11 (1996). https://doi.org/10.1016/1350-4533(95)00024-0
32. B. Pesaran, Spectral analysis for neural signals, in P. Mitra (ed.) *Neural Signal Processing: Quantitative Analysis of Neural Activity. Society for Neuroscience* (2008), pp. 1–12
33. K.M.M. Prabhu, V.U. Reddy, J.P. Agrawal, Performance comparison of data windows. Electron. Lett. **13**, 600–601 (1977). https://doi.org/10.1049/el:19770431
34. J. Proakis, D. Manolakis, *Digital Signal Processing*, 4th ed. (Prentice Hall, 2006). Doi: citeulike-article-id:1806381
35. S. Sanei, J.A. Chambers, S. Sanei, J.A. Chambers, Fundamentals of EEG Signal Processing, in *EEG Signal Processing*. (Wiley, New York, 2007), pp. 35–125. https://doi.org/10.1002/9780470511923.ch2
36. C.E. Shannon, Communication in the presence of noise. Proc. IRE **37**, 10–21 (1949). https://doi.org/10.1109/jrproc.1949.232969
37. N.S. Singhal, J.E. Sullivan, Continuous spike-wave during slow wave sleep and related conditions ISRN. Neurology **2014**, 619079 (2014). https://doi.org/10.1155/2014/619079
38. S.J.M. Smith, EEG in the diagnosis, classification, and management of patients with epilepsy. J. Neurol. Neurosurg. Psychiatry **76**, ii2–ii7 (2005). https://doi.org/10.1136/jnnp.2005.069245
39. T. Takahashi et al., Changes in EEG and autonomic nervous activity during meditation and their association with personality traits. Int. J. Psychophysiol. **55**, 199–207 (2005)
40. D.J. Thomson, Spectrum estimation and harmonic analysis. Proc. IEEE **70**, 1055–1096 (1982). https://doi.org/10.1109/PROC.1982.12433

41. J. Tzyy-Ping, S. Makeig, M. Stensmo, T.J. Sejnowski, Estimating alertness from the EEG power spectrum. IEEE Trans. Biomed. Eng. **44**, 60–69 (1997). https://doi.org/10.1109/10.553713
42. P. Welch, The use of fast Fourier transform for the estimation of power spectra: a method based on time averaging over short, modified periodograms. IEEE Trans. Audio Electroacoust. **15**, 70–73 (1967). https://doi.org/10.1109/TAU.1967.1161901

Chapter 4
The Analysis of Event-Related Potentials

Marco Congedo

Abstract In this chapter, we provide an introduction to the major methods used for the analysis and classification of Event-Related Potentials (ERPs). We start by considering the problem of estimating ERP ensemble averages in the time domain. An estimator allowing for weights and time shifts for each trial is discussed. Then we consider spatial, temporal and spatio-temporal multivariate filters for improving the estimation, including principal component analysis, the common spatial pattern and blind source separation. Then, we review time-frequency analysis methods. The reader is provided with definitions in order to understand the most commonly used linear and non-linear measures used in the time-frequency domain. We continue with a brief discussion on the importance of the analysis in the spatial domain, including topographic maps and tomographies. Next, we review procedures for applying inferential statistics to ERP studies. Emphasis is given to procedures based on permutation tests, which account for the multiple comparison problem and adapt to the form and degree of correlation between hypotheses. Finally, we consider the problem of classifying ERP single-trials, pointing to recent literature covering the most promising methods currently available, namely, Riemannian geometry, random forests and neural networks.

4.1 Introduction

Event-Related Potentials (ERPs) are a fundamental class of phenomena that can be observed by means of electroencephalography (EEG). They are defined as potential difference fluctuations that are both time-locked and phase-locked to a discrete physical, mental, or physiological occurrence, referred to as the *event*. ERPs are usually described as a number of positive and negative peaks characterized by their polarity, shape, amplitude, latency and spatial distribution on the scalp. All these

M. Congedo (✉)
GIPSA-Lab, Centre National de la Recherche Scientifique (CNRS), Grenoble-INP, Université Grenoble Alpes, Grenoble, France
e-mail: marco.congedo@gmail.com

C.-H. Im (ed.), *Computational EEG Analysis*, Biological and Medical Physics, Biomedical Engineering, https://doi.org/10.1007/978-981-13-0908-3_4

characteristics depend on the type (class) of event. Each realization of an ERP is named a *sweep* or trial. Important pioneering discoveries of ERPs include the contingent negative variation [96], the P300 [88], the mismatch negativity [68] and the error-related negativity [31]. Another class of time-locked phenomena are the Event-Related De/Synchronizations (ERDs/ERSs, [77]), which are *not* phase-locked. In order to keep a clear distinction between the two, ERD/ERS are referred to as *induced* phenomena, while ERPs are referred to as *evoked* phenomena [89]. Traditionally, ERPs have been conceived as stereotypical fluctuations with approximately fixed polarity, shape, latency, amplitude and spatial distribution. Accordingly, the ERP fluctuations are independent from the ongoing EEG and superimpose to it in a time- and phase-locked fashion with respect to the triggering event. This yields the so-called *additive* generative model. Several observations have challenged this model [23], suggesting the possibility that evoked responses may be caused by a process of *phase resetting*, that is, an alignment of the phase of the spontaneous neuronal activity with respect to the event [44, 57, 62]. According to this model, ERPs result from time/frequency modulations of the ongoing activity of specific neuronal populations. Still another generative model of ERPs was introduced by [65] and [69]. These authors pointed out that ongoing EEG activity is commonly non-symmetric around zero, as can be seen clearly in sub-dural recordings of alpha rhythms [58]. They proposed that averaging amplitude-asymmetric oscillations may create evoked responses with slow components.

In this chapter, we consider several major methods currently used to analyze and classify ERPs. In modern EEG, using a multitude of electrodes is the rule rather than the exception, thus emphasis is given on multivariate methods, since these methods can exploit spatial information and achieve higher signal-to-noise ratio (SNR) as compared to single-electrode recordings. We consider the analysis in the time domain, in the time-frequency domain and in the spatial domain. We also consider inter-trial amplitude and latency variability as well as the case of overlapping ERPs. We then consider useful tools for inferential statistics and classifiers for machine learning specifically targeting ERP data. All the time-domain methods described in this chapter are implicitly based on the additive model, but they may give meaningful results even if the data is generated under other models. Time-frequency domain methods can explicitly study the phase consistency of ERP components. We will show an example analysis for each section. The real data examples in all but the last figure concerns a visual P300 experiments where healthy adults play a brain-computer interface video-game named *Brain Invaders* [21]. This experiment is based on the classical oddball paradigm and yields ERPs pertaining to a *target class*, evoked by infrequent stimuli, and a *non-target class*, evoked by frequent stimuli.

4.2 General Considerations in ERP Analysis

ERP analysis is always preceded by a pre-processing step in which the data is digitally filtered. Notch filters for suppressing power line contamination and band-pass filters are common practice to increase the SNR and remove the direct current level [61]. If the high-pass margin of the filter is lower than 0.5 Hz, the direct current level can be eliminated by subtracting the average potential (baseline) computed on a short window before the ERP onset (typically 250 ms long). Researchers and clinicians are often unaware of the signal changes that can be introduced by a digital signal filter, yet the care injected in this pre-processing stage is well rewarded, since severe distortion in signal shape, amplitude, latency and even scalp distribution can be introduced by an inappropriate choice of digital filter [98].

There is consensus today that for a given class of ERPs only the polarities of the peaks may be considered consistent for a given electrical reference used in the EEG recording; the shape, latency, amplitude and spatial distribution of ERPs are highly variable among individuals. Furthermore, even if within each individual the shape may be assumed stable on average, there may be a non-negligible amplitude and latency inter-sweep variability. Furthermore, the spatial distribution can be considered stable within the same individual and within a recording session, but may vary from session to session, for instance, due to slight differences in electrode positioning. Inter-sweep variability is caused by the combination of several experimental, biological and instrumental factors. Experimental and biological factors may affect both latency and amplitude. Examples of experimental factors are the stimulus intensity and the number of items in a visual search task [61]. Examples of biological factors are the subject's fatigue, attention, vigilance, boredom and habituation to the stimulus. Instrumental factors mainly affect the latency variability; the ERP marking on the EEG recording may introduce a jitter, which may be non-negligible if the marker is not recorded directly on the EEG amplification unit and appropriately synchronized therein, or if the stimulation device features a variable stimulus delivery delay. An important factor of amplitude variability is the ongoing EEG signal; large artifacts and high energy background EEG (such as the posterior dominant rhythm) may affect differently the sweeps, depending on their amplitude and phase, artificially enhancing or suppressing ERP peaks.

Special care in ERP analysis must be undertaken when we record overlapping ERPs, since in this case simple averaging results in biased estimations [85, 99, 100]. ERPs are non-overlapping if the minimum inter-stimulus interval (ISI) is longer than the length of the latest recordable ERP. There is today increasing interest in paradigms eliciting overlapping ERPs, such as some odd-ball paradigms [21] and rapid image triage [104], which are heavily employed in brain-computer interfaces for increasing the information transfer rate [101] and in the study of eye-fixation potentials, where the "stimulus onset" is the time of an eye fixation and saccades follow rapidly [86]. The strongest distortion is observed when the ISI is fixed. Less severe is the distortion when the ISI is drawn at random from an exponential distribution [21, 85].

Amplitude/latency inter-sweep variability as well as the occurrence of overlapping ERPs call for specific analysis methods. In general, such methods result in an improved ensemble average estimation. For a review of such methods, the reader is referred to Congedo and Lopes da Silva [23].

4.3 Time Domain Analysis

The main goal of the analysis in the time domain is to estimate the *ensemble average* of several sweeps and characterize the ERP peaks in terms of amplitude, shape and latency. Using matrix algebra notation, we will denote by $x(t)$, the column vector holding the multivariate EEG recording at N electrodes and at time sample t, whereas $\mathrm{N} \times \mathrm{T}$ matrix X_k will denote a data epoch holding the kth observed sweep for a given class of ERP signals. These sweeps last T samples and start at event time $\pm$ an offset that depends on the ERP class. For instance, the ERPs and ERDs/ERSs follow a visual presentation but precede a button press. The sweep onset must therefore be set accordingly adjusting the offset. We will assume along this chapter that T>N, i.e., that the sweeps comprise more samples than sensors. We will index the sweeps for a given class by $k \in \{1, \ldots, \mathrm{K}\}$, where K is the number of available sweeps for the class under analysis.

4.3.1 The Additive Generative Model

The *additive* generative model for the observed sweep of a given class can be written as

$$X_k = \sigma_k Q(\tau_k) + N_k, \tag{4.1}$$

where Q is an $\mathrm{N} \times \mathrm{T}$ matrix representing the stereotypical evoked responses for the class under analysis, σ_k are positive scaling factors accounting for inter-sweep variations in the amplitude of Q, τ_k are time-shifts, in samples units, accounting for inter-sweep variations in the latency of Q and N_k are $\mathrm{N} \times \mathrm{T}$ matrices representing the noise term added to the kth sweep. Here by 'noise' we refer to all non-evoked activity, including ongoing and induced activity, plus all artifacts. According to this model, the evoked response in Q is continuously modulated in amplitude and latency across sweeps by the aforementioned instrumental, experimental and biological factors. Therefore, the single-sweep SNR is the ratio between the variance of σ_k $Q(\tau_k)$ and the variance of N_k. Since the amplitude of ERP responses on the average is in the order of a few μV, whereas the noise is in the order of several tens of μV, the SNR of single sweeps is very low. The classical way to improve the SNR is averaging several sweeps. This enhances evoked fluctuations by constructive interference, since they are the only time- and phase-locked fluctuations.

4.3.2 Ensemble Average Estimations

The usual *arithmetic ensemble average* of the K sweeps is given by

$$\bar{X} = \frac{1}{K}\sum_{k=1}^{K} X_k. \tag{4.2}$$

This estimator is unbiased if the noise term is zero-mean, uncorrelated to the signal, spatially and temporally uncorrelated and stationary. It is actually optimal if the noise is also Gaussian [56]. However, these conditions are never matched in practice. For instance, EEG data are both spatially and temporally correlated and typically contain outliers and artifacts, thus are highly non-stationary. As a rule of thumb, the SNR of the arithmetic ensemble average improves proportionally to the square root of the number of sweeps. In practice, it is well known that the arithmetic mean is an acceptable ensemble average estimator provided that sweeps with low SNR are removed and that enough sweeps are available. A better estimate is obtained by estimating the weights σ_k and shift τ_k to be given to each sweep before averaging. The resulting *weighted and aligned arithmetic ensemble average* is given by

$$\bar{X} = \frac{\sum_{k=1}^{K} (\sigma_k X_k(\tau_k))}{\sum_{k=1}^{K} \sigma_k}. \tag{4.3}$$

Of course, with all weights equal and all time-shifts equal to zero, ensemble average estimation (4.3) reduces to (4.2). Importantly, when ERP overlaps, as discussed above, estimators (4.2) or (4.3) should be replaced by a multivariate regression version, which is given by (1.9) in Congedo et al. [22].

4.3.3 Multivariate Filtering Methods

A large family of multivariate methods have been developed with the aim of improving the estimation of ERP ensemble averages by means of spatial, temporal or spatio-temporal filtering. These filters transform the original time-series of the ensemble average in a number of components, which are linear combinations of the original data. A spatial filter outputs components in the form of time-series, which are linear combinations of sensors for each sample, along with the spatial patterns corresponding to each component. A temporal filter outputs components in the form of spatial maps, which are linear combinations of samples for each sensor, along with the temporal patterns corresponding to each component. A spatio-temporal filter outputs components that are linear combinations of sensor and samples at the same time, along with the corresponding spatial and temporal patterns. Given an ensemble average estimation such as in (4.2) or (4.3), the output of the spatial, temporal, and spatio-temporal filters are the components given by

$$\begin{cases} \bar{Y} = B^T \bar{X} & \text{spatial} \\ \bar{Y} = \bar{X} D & \text{temporal} \\ \bar{Y} = B^T \bar{X} D & \text{spatio - temporal} \end{cases} . \tag{4.4}$$

For both the N × P spatial filter matrix B and the T × P temporal filter matrix D, we require 0<P<N, where P is named the *subspace dimension*. The upper bound for P is due to the fact that for our data N<T and that filtering is achieved effectively by discarding from the ensemble average the N-P components not accounted for by the filters, that is, at least one component must be discarded. The task of a filter is indeed to decompose the data in a small number of meaningful components so as to suppress noise while enhancing the relevant signal. Once designed the matrices B and/or D, the filtered ensemble average estimation is obtained by projecting back the components onto the sensor space, as

$$\begin{cases} \bar{X}' = AB^T \bar{X} & \text{spatial} \\ \bar{X}' = \bar{X} DE^T & \text{temporal} \\ \bar{X}' = AB^T \bar{X} DE^T & \text{spatio - temporal} \end{cases} , \tag{4.5}$$

where N × P matrix A and T × P matrix E are readily found so as to verify

$$B^T A = E^T D = I. \tag{4.6}$$

In the spatio-temporal setting the columns of matrix A and E are the aforementioned spatial and temporal patterns, respectively. In the spatial setting, only the spatial patterns in A are available, however the components in the rows of $\bar{Y}$(spatial) in (4.4) will play the role of the temporal patterns. Similarly, in the temporal setting, only the temporal patterns in E are available, however the components in the columns of $\bar{Y}$(temporal) in (4.4) will play the role of the spatial patterns. So, regardless the type of chosen filter, in this kind of analysis it is customary to visualize the spatial patterns in the form of scalp topographic or tomographic maps and the temporal pattern in the form of associated time-series. This way one can evaluate the spatial and/or temporal patterns of the components that should be retained and those that should be discarded so as to increase the SNR. Nonetheless, we stress here that in general these patterns bear no physiological meaning. A notable exception are the patterns found by the family of blind source separation methods, discussed below, which, under a number of assumptions, allow such interpretation.

4.3.4 Principal Component Analysis

Principal component analysis (PCA) has been the first multivariate filter of this kind to be applied to ERP data [28, 45] and has been often employed [12, 27, 51]. A long-lasting debate has concerned the choice of the spatial vs. temporal PCA [27, 79], however we hold here that the debate is resolved by performing a spatio-temporal PCA, combining the advantages of both. The PCA seeks uncorrelated components maximizing the variance of the ensemble average estimation (4.2) or (4.3); the first component explains the maximum of its variance, while the remaining components explain the maximum of its remaining variance, subjected to being uncorrelated to all the previous. Hence, the variance explained by the N-P discarded components explains the variance of the 'noise' that has been filtered out by the PCA. In symbols, the PCA seeks matrices B and/or D with orthogonal columns so as to maximize the variance of $\bar{X}'$. Note that for any choice of 0<P<N, the filtered ensemble average estimator $\bar{X}'$ obtained by PCA is the best P-rank approximation to $\bar{X}$ in the least-squares sense, i.e., for any 0<P<N, the matrices B and/or D as found by PCA attain the minimum variance of $\bar{X} - \bar{X}'$.

The PCA is obtained as it follows: let

$$\bar{X} = UWV^T \tag{4.7}$$

be the singular-value decomposition of the ensemble average estimation, where N × T matrix W holds along the principal diagonal the N non-null singular values in decreasing order ($w_1 \geq \cdots \geq w_N$) and where N × N matrix U and T × T matrix V hold in their columns the left and right singular vectors, respectively. Note that the columns of U and V are also the eigenvectors of $\bar{X}\bar{X}^T$ and $\bar{X}^T\bar{X}$, respectively, with corresponding eigenvalues in both cases being the square of the singular values in W and summing to the variance of $\bar{X}'$. The spatial PCA is obtained filling B with the first P column vectors of U, the temporal PCA is obtained filling D with the first P column vectors of V and the spatio-temporal PCA is obtained filling them both. The appropriate version of (4.4) and (4.5) then applies to obtain the components and the sought filtered ensemble average estimation, respectively. In all cases 0<P<N is the chosen subspace dimension. Note that since for PCA the vectors of the spatial and/or temporal filter matrix are all pair-wise orthogonal, (4.6) is simply verified by setting $A = B$ and/or $E = D$.

An example of spatio-temporal PCA applied to an ERP data set is shown in Fig. 4.1, using estimator (4.2) in the second column and estimator (4.3) in the fourth column. The ERP of this subject features a typical N1/P2 complex at occipital locations and an oscillatory process from about 50–450 ms, better visible at central and parietal location, ending with a large positivity peaking at 375 ms (the "P300"). We see that by means of only four components the PCA effectively compresses the ERP, retaining the relevant signal; however, eye-related artefacts are also retained (see traces at electrodes FP1 and FP2). This happens because the variance of these artefacts is very high, thus as long as the artefacts are somehow spatially and temporally consistent across sweeps, they will be retained in early components along with the consistent (time and phase-locked) ERPs, even if estimator (4.3) is used. For this reason, artefact rejection is generally necessary before applying a PCA.

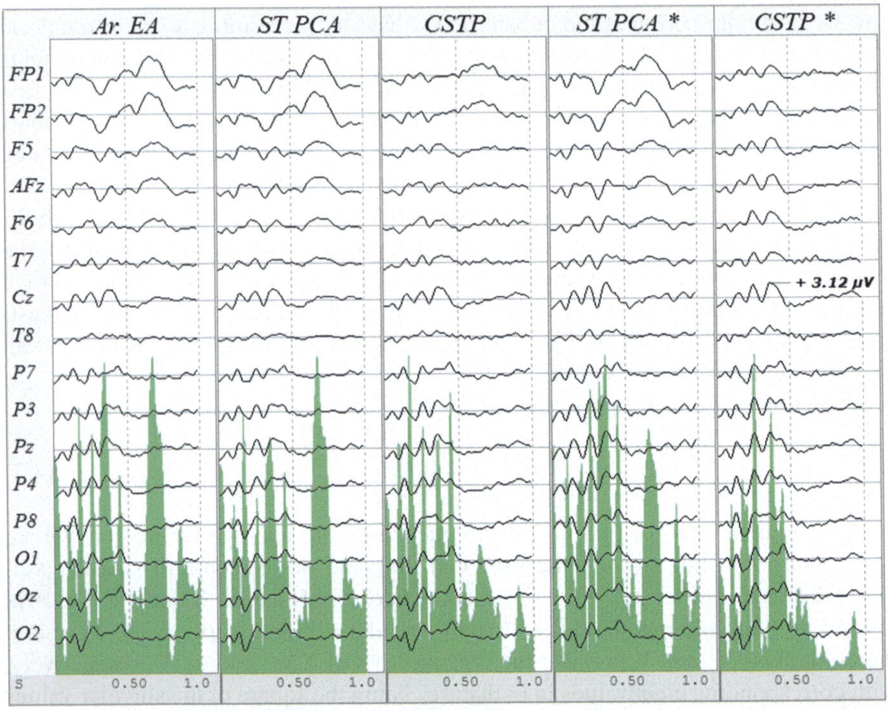

Fig. 4.1 Comparison of several filtered ensemble average estimations via (4.5) using several spatio-temporal filtering methods. One second of data starting at target (infrequent stimulus) presentation averaged across 80 sweeps is displayed. No artifact rejection was performed. The green shaded area is the global field power (Lehmann and Skrandies [52] in arbitrary units, *Legend* "Ar. EA" = non-filtered arithmetic mean ensemble average given by (4.2). "ST PCA" = spatio-temporal PCA with P = 4. "CSTP" = CSTP with P = 12; These two filters have been applied to estimator (4.2). "*" The filters are applied on the weighted and aligned estimator (4.3) using the adaptive method of Congedo et al. [22]. All plots have the same horizontal and vertical scales

4.3.5 The Common Pattern

In order to improve upon the PCA we need to define a measure of the SNR, so that we can devise a filter maximizing the variance of the evoked signal, like PCA does, while also minimizing the variance of the noise. Consider the average spatial and temporal sample covariance matrix when the average is computed across all available sweeps, such as

$$S = \frac{1}{K}\sum_{k=1}^{K} COV(X_k), \quad T = \frac{1}{K}\sum_{k=1}^{K} COV\left(X_k^T\right) \tag{4.8}$$

and the covariance matrices of the ensemble averages, namely,

$$\bar{S} = COV(\bar{X}),\ \bar{T} = COV(\bar{X}^T). \tag{4.9}$$

The quantities in (4.8) and (4.9) are very different; in fact S and T hold the covariance of all EEG processes that are active during the sweeps, regardless the fact they are time and phase-locked or not, while in $\bar{S}$ and $\bar{T}$ the non-phase-locked signals have been attenuated by computing the ensemble average in the time domain. That is to say, referring to model (4.1), S and T contain the covariance of the signal plus the covariance of the noise, whereas $\bar{S}$ and $\bar{T}$ contain the covariance of the signal plus an attenuated covariance of the noise. A useful definition of the SNR for the filtered ensemble average estimation is then

$$\mathrm{SNR}(\bar{X}') = \frac{\mathrm{VAR}(AB^T\bar{X}DE^T)}{\frac{1}{\mathrm{K}}\sum_{k=1}^{\mathrm{K}}\mathrm{VAR}(AB^T\bar{X}_kDE^T)}. \tag{4.10}$$

The common spatio-temporal pattern (CSTP), presented in Congedo et al. [22], is the filtering method maximizing this SNR. It can be used as well when the data contains several classes of ERPs. The sole spatial or temporal common pattern approaches are obtained as special cases. Both conceptually and algorithmically, the CSTP can be understood as a PCA performed on whitened data. So, the PCA can be obtained as a special case of the CSTP by omitting the whitening step. The reader is referred to Congedo et al. [22] for all details and reference to available code libraries. An example of CSTP is shown in Fig. 4.1. In contrast to the spatio-temporal PCA, the CSTP has removed almost completely the eye-related artefact. The last two plots in Fig. 4.1 show the filtered ensemble average estimation obtained by spatio-temporal PCA and CSTP using the adaptive method presented in Congedo et al. [22] for estimating the weights and shift so as to use (4.3) instead of (4.2); the CSTP estimator is even better in this case, as residual eye-related artefacts at electrodes FP1 and FP2 have been completely eliminated.

4.3.6 Blind Source Separation

Over the past 30 years, *Blind Source Separation* (BSS) has established itself as a core methodology for the analysis of data in a very large spectrum of engineering applications such as speech, image, satellite, radar, sonar, antennas and biological signal analysis [17]. In EEG, BSS is often employed for denoising/artifact rejection (e.g., [26]) and in the analysis of continuously recorded EEG, ERDs/ERSs and ERPs. Traditionally, BSS operates by spatially filtering the data. Therefore, it can be casted out in the framework of spatial filters we have previously presented, that is, using the first of the three expressions in (4.4) and (4.5). We have seen that PCA and the common pattern filter seek abstract components optimizing some criterion: the signal variance for PCA and an SNR for the common pattern. In contrast, BSS aims at estimating the *true brain dipolar components* resulting in the observed scalp measurement. For doing so, BSS makes a number of assumptions. The common one for all BSS methods is that the observed EEG potential results from an instantaneous linear mixing of a number of cortical dipolar electric fields. Although this is an approximation of the physical process of current generation in the brain and diffusion through the head [70], physical and physiological knowledge support such generative model for scalp potentials [9]. In particular, the model fits well low-frequency electrical phenomena with low spatial resolution, which yield the strongest contribution to the recordable EEG. The model reads

$$x(t) = As(t), \tag{4.11}$$

where, as before, *x(t)* is the observed N-dimensional sensor measurement vector, *s(t)* the unknown P-dimensional vector holding the true dipolar *source process* (with $0<P\leq N$), and *A*, also assumed unknown in BSS, is named the *mixing matrix*. BSS entails the estimation of a *demixing matrix B* allowing source process estimation

$$y(t) = B^T x(t). \tag{4.12}$$

We say that the source process can be *identified* if

$$y(t) \approx Gs(t), \tag{4.13}$$

where $P \times P$ matrix $G = B^T A$ is a scaled permutation matrix, i.e., a square matrix with only one non-null element in each row and each column. Matrix *G* cannot be observed since *A* is unknown. It enforces a shuffling of the order and amplitude (including possible sign switching) of the estimated source components, which cannot be solved by BSS. Equation (4.13) means that in BSS the actual *waveform* of the source process has been approximately identified, albeit the sign, scaling and order of the estimated source process is arbitrary. Such identification is named *blind* because no knowledge on the source waveform *s*(*t*) nor on the mixing process *A* is assumed. Fortunately, condition (4.13) can be achieved under some additional assumptions relating to the statistical properties of the dipolar source components (see [11, 78]).

Two important families of BSS methods operate by canceling inter-sensor second order statistics (SOS) or higher (than two) order statistics (HOS); the latter family being better known as *independent component analysis* (ICA) (see [17], for an overview). In doing so, both assume some form of *independence among the source processes*, which is specified by inter-sensor statistics that are estimated from the data. The difference between the two families resides in the assumption about the nature of the source process; since Gaussian processes are defined exhaustively by their mean and variance (SOS), ICA may succeed only when at most one of the components is Gaussian. On the other hand, SOS methods can identify the source process components regardless of their distribution, i.e., even if they are all Gaussian, but source components must have a unique power spectrum signature and/or a unique pattern of energy variation across time, across experimental conditions or, in the case of ERPs, across ERP classes (see [20, 24]). For HOS methods the available EEG can be used directly as input of the algorithms [26]. For BSS methods, either lagged covariance matrices or Fourier co-spectral matrices are estimated on the available data, then the demixing matrix B is estimated as the approximate joint diagonalizer of all these matrices [20]. Details on the application of BSS methods to ERP data can be found in Congedo et al. [24].

Figure 4.2 shows the result of a SOS-based BSS analysis applied to P300 data; here the ensemble averages have been aligned using the method described by Congedo et al. [22]. Analyzing both the temporal course and spatial distribution, we see that the BSS analysis finds two relevant source components: S7 features a topographic map (spatial pattern) with maximum at the vertex and an ERP (temporal pattern) with maximum at 370 ms, clearly describing the P300. S13 features a topographic map with maximum at parietal and occipital bilateral derivations and an ERP with the classical P100/N200 complex describing a visual ERP. Both source components are present only in the target sweeps. Further analysis of these components will be presented in the Sect. 4.4. time-frequency domain analysis. Clearly, BSS has successfully separated the two ERP components.

It is worth mentioning that while traditionally only spatial BSS is performed, a spatio-temporal BSS method for ERPs has been presented in Korczowski et al. [49]. Just as in the case of PCA and common pattern, a spatio-temporal approach is preferable for ERP analysis, thus it should be pursued further (Fig. 4.2).

4.4 Time-Frequency Domain Analysis

Time-Frequency Analysis (TFA) complements and expands the time domain analysis of ERP thanks to a number of unique features. While the analysis in the time domain allows the study of phase-locked ERP components only, TFA allows the study of both phase-locked (evoked) and non-phase-locked (induced) ERP components. In addition to timing, the TFA provides information about the *frequency* (both for evoked and induced components) and about the *phase* (evoked components only) of the underlying physiological processes. This is true for the analysis of a single time series (univariate) as well as for the analysis of the dependency between two time-series (bivariate), the latter not being treated here. In all cases, the time series under analysis may be the sweeps derived at significant scalp derivations or BSS source components with specific physiological meaning as obtained by the methods we have discussed above. In this section we introduce several univariate TFA measures.

A time-frequency analysis (TFA) decomposes a signal in a two dimensional plane, with one dimension being the *time* and the other being the *frequency*. Whereas several possible time-frequency representations exist, nowadays in ERP studies we mainly encounter *wavelets* [50, 89] or the *analytic signal* resulting from the *Hilbert transform* [13, 84, 90]. Several studies comparing wavelets and the Hilbert transform have found that the two representations give similar results [8, 53].

The example we provide below employs the Hilbert transform [37], which is easily and efficiently computed by means of the fast Fourier transform [64]. By applying

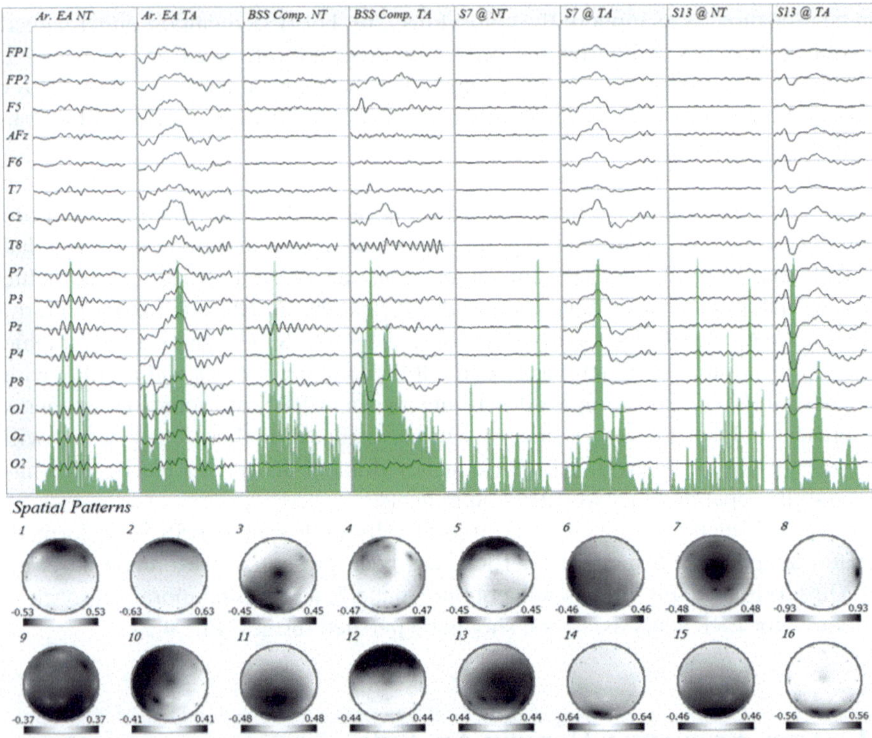

Fig. 4.2 SOS-based blind source separation of ERP. From left to right of the top panel: the weighted and aligned ensemble average (4.3) of the non-target sweeps (Ar. EA NT) and of the target sweeps (Ar. EA TA), the BSS components for non-target (BSS Comp. NT) and target (BSS Comp. TA) ensemble average (obtained via (4.4), first expression), the same filtered ensemble average retaining source component 7 for the non-target (S7 @ NT) and target sweeps (S7 @ TA) and the filtered ensemble average obtained retaining source component 13 for the non-target (S13 @NT) and target sweeps (S13 @ TA). +: arbitrary vertical units for each trace. The bottom panel shows the spatial patterns (columns of the inverse of matrix *B*) of the BSS components in the form of monochromatic topographic maps. The sign of the potential is arbitrary in BSS analysis. Each map is scaled to its own maximum. Note the separation of two source components: S7 which accounts for the P300, with maximum at the vertex and an ERP with maximum at 370 ms, and S13, which accounts for the classic P100/N200 visual ERP, with maximum at parietal and occipital bilateral derivations. As expected, both source components are present only in the target sweeps, whereas other components are visible in both the target and non-target sweeps

a filter bank to the signal, that is, a series of band-pass filters centered at successive frequencies f (for example, centered at 1 Hz, 2 Hz, …) and by computing the Hilbert transform for each filtered signal, we obtain the analytic signal in the time-frequency representation. Each time-frequency point of the *analytic signal* is a complex number $z_{tf}=a_{tf}+ib_{tf}$ (Fig. 4.3). For each sample of the original signal we obtain from z_{tf} the *instantaneous amplitude* r_{tf}, also known as the *envelope*, as its *modulus* $r_{tf}=|z_{tf}|$ and the *instantaneous phase* φ_{tf} as its *argument* $\varphi_{tf}=\mathrm{Arg}\ (z_{tf})$. The amplitude r_{tf}

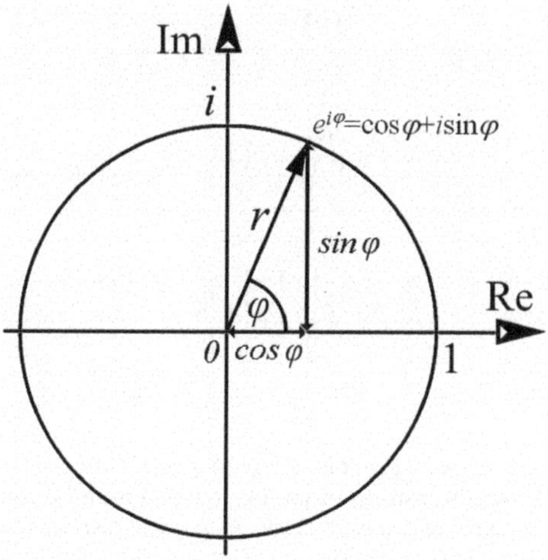

Fig. 4.3 In the complex plane the abscissa is the real line and the ordinate is the imaginary line endowed with the imaginary unit i, which is defined as $i^2 = -1$. A complex number can be represented in Cartesian form as the point $z = a + ib$ in such plane, where a is the real coordinate and ib is the imaginary coordinate. The point can be represented also by a position vector, that is, the vector joining the origin and the point, with length r and angle φ (in the left part of the figure the point is on the unit circle). r and φ are known as the polar coordinates. In trigonometric form the coordinates are $r\cos\varphi$ and $ir\sin\varphi$, therefore, using Euler's formula $e^i = \cos\varphi + i\sin\varphi$, we can also express any complex number as $z = re^i$

is expressed in μV units. The phase φ_{tf} is a cyclic quantity usually reported in the interval $(-\pi, \ldots, \pi]$, but can be equivalently reported in any interval such as $(-1, \ldots, 1]$, $(0, \ldots, 1]$ or in degrees $(0°, \ldots, 360°]$. The physical meaning and interpretation of the analytic signal, the instantaneous amplitude and the instantaneous phase are illustrated in Fig. 4.4. Besides illustrating these concepts, the simple examples in Fig. 4.4 shows how prone to errors may be the interpretation of the analytic signal if a filter bank is not used.

There are two ways of averaging the analytic signal across sweeps. The first is sensitive to evoked (phase-locked) ERP components. The second is sensitive to both evoked and induced (non-phase-locked) components. Thus, we obtain complementary information using the two averaging procedures. In order to study evoked components we average directly the analytic signal at each time-frequency point, such as

$$\bar{z}_{tf} = \frac{1}{K}\sum_k a_{ktf} + i\frac{1}{K}\sum_k b_{ktf} \tag{4.14}$$

from which the *average instantaneous amplitude* (envelope) is given by

$$\bar{r}_{ft} = \left|\bar{z}_{tf}\right| \tag{4.15}$$

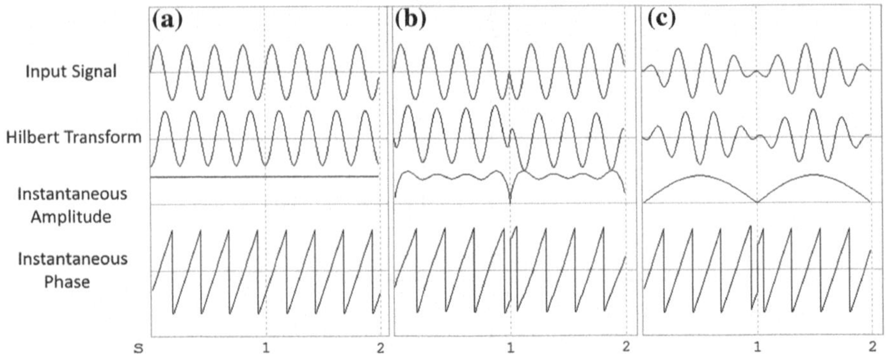

Fig. 4.4 Three 2-second signals were generated (input Signal). Time is on the abscissa. The vertical scaling is arbitrary. The Hilbert transform of the input signal is shown in the second traces. The next two traces are the instantaneous amplitude (envelope) and instantaneous phase. Note that the envelope is a non-negative quantity. **a** The input signal is a sine wave at 4 Hz. The instantaneous amplitude is constant in the whole epoch. The phase oscillates regularly in between its bounds at 4 Hz. **b** The input signal is a sine wave at 4 Hz with a phase discontinuity occurring exactly in the middle of the epoch. The instantaneous amplitude now drops in the middle of the epoch. As expected, the instantaneous phase features a discontinuity in the middle of the epoch. **c** the input signal is a sine wave at 4 Hz multiplied by a sine wave at 0.5 Hz with the same amplitude. The result input signal is a sine wave at 4 Hz, which amplitude and phase are modulated by the sine wave at 0.5 Hz. The Instantaneous amplitude is the envelope of the sine at 0.5 Hz. The instantaneous phase is like the one in B, but is now caused by the multiplication with the 0.5 Hz wave

and the *average instantaneous phase* is given by

$$\bar{\varphi}_{tf} = \arg\left(\bar{z}_{tf}\right) \tag{4.16}$$

Note that in this case the envelope may be high only if the sweeps at that time-frequency point have a *preferred phase*, whereas if the phase is randomly distributed from sweep to sweep, the average envelope will tend toward zero. This phenomenon is illustrated in Fig. 4.5.

While the Hilbert transform is a linear operator, *non-linear* versions of measures (4.15) and (4.16) may be obtained by adding a simple *normalization* of the analytic signal at each sweep [74]; before computing the average in (4.14), replace a_{ktf} by a_{ktf} / r_{ktf} and b_{ktf} by b_{ktf} / r_{ktf}, where $r_{ktf} = \sqrt{a_{ktf}^2 + b_{ktf}^2}$ is the modulus. This means that at all time-frequency points and for each sweep the complex vector $a_{ktf} + ib_{ktf}$ is stretched or contracted so as to be constrained on the unit complex circle (Fig. 4.6). The average instantaneous amplitude (4.15) and phase (4.16) after the normalization will be actually sensitive to the stability of the phase across sweeps, *regardless of amplitude*. Such non-linear measure is known as *inter-trial phase coherence* (ITPC: [62]), but has been named by different authors also as "inter-trial phase clustering", "phase coherence" among other ways [14].

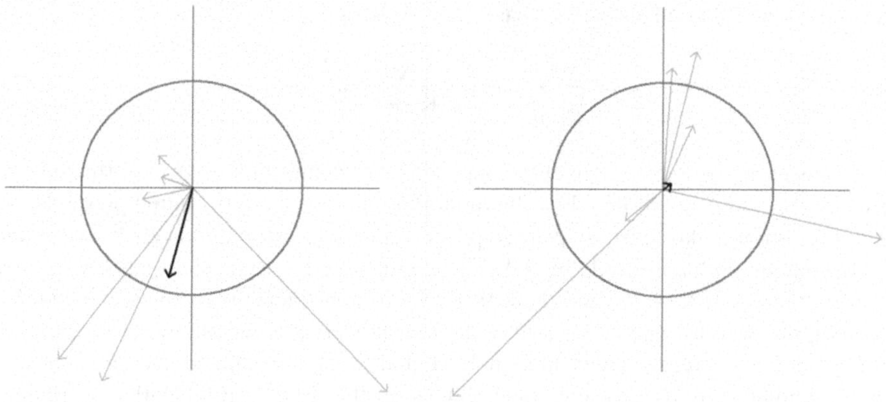

Fig. 4.5 In each diagram six complex numbers are represented as position vectors (gray arrows) in the complex plane (see Fig. 4.3). Consider these vectors as representing the analytic signal for a given time-frequency point estimated on six different sweeps. In each diagram the black arrow is the position vector corresponding to the average of the six complex numbers as per (4.15). In the left diagram the vectors are distributed within one half circle, featuring a preferred direction. In the right diagram the vectors are more randomly distributed around the circle; the resulting mean vector is much smaller, although the average length of the six vectors in the two diagram is approximately equal

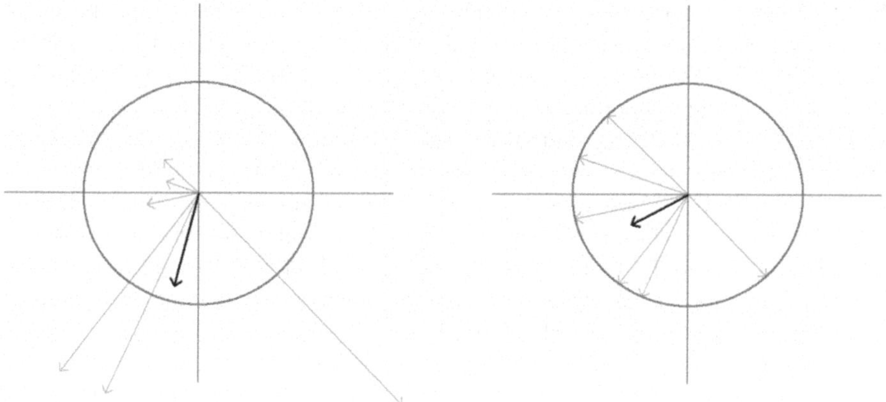

Fig. 4.6 The left diagram is the same as in Fig. 4.5. The vectors in the right diagram have been normalized to unit length (non-linear normalization). Note that the mean vector on the right points in a different direction as compared to the mean vector on the left, albeit the vectors have the same direction in the two diagrams; while on the left diagram the amplitude of the vectors weights the average, on the right diagram the amplitude is ignored

If induced components are of interest, instead of using (4.14) we average the envelope computed on each sweep as

$$\bar{r}_{tf} = \frac{1}{K}\sum_{k}|z_{ktf}| = \frac{1}{K}\sum_{k}\sqrt{a_{ktf}^{2} + b_{ktf}^{2}} \tag{4.17}$$

In this case, the average envelope depends on the amplitude of the coefficients in each sweep and is not affected by the randomness of the analytic signal phase. Note that it does not make sense to average phase values φ_{ktf} estimated at each sweep, as we have done with amplitude in (4.17), since the phase is a circular quantity.[1]

Measures (4.15), (4.16) and their normalized (non-linear) versions can be modified computing a *weighted average* of the normalized analytic signal. Note that the non-normalized average analytic signal is equal to the normalized average analytic signal weighted by its own envelope. Choosing the weights differently, we obtain quite different measures of phase consistency. For instance, weights can be given by experimental or behavioral variables such as reaction time, stimulus luminance, etc. In this way, we can discover phase consistency effects that are specific to certain properties of the stimulus or certain behavioral responses [14, 15]. Taking as weight the envelope of the signal at the frequency under analysis and the analytic signal of *another frequency* (that we name here the *modulating* frequency) we obtain a measure of phase-amplitude coupling named *modulation index* (MI: [10, 14, p. 413]). If the distribution of the modulating phase is uniform, high values of MI reveal dependency between the two frequencies. The modulating frequency is usually lower than the frequency under analysis. Note that by weighting the normalized analytic signal arbitrarily, the obtained average amplitude is no longer guaranteed to be bounded superiorly by 1.0. Furthermore, such measures are subjected to several confounding effects and must be standardized using resampling methods (for details see [10, 14, pp. 253–257, 413–418]). An alternative to the MI measure that does not require such standardization is the *phase-amplitude coupling* (PAC), which is the MI normalized by the amplitude [72]. Measures such as MI and PAC and other variants, along with bivariate counterparts (e.g., [94]), are used to study an important class of phenomena that can be found in the literature under the name of amplitude-amplitude, phase-amplitude and phase-phase nesting (or coupling, interaction, binding…), amplitude modulation and more [16, 36, 54, 55, 73, 93].

Several measures of amplitude and phase in the time-frequency plane are shown in the following real-data example. Figure 4.7 shows a time-frequency analysis of source S7 and S13 of Fig. 4.2. The analysis has been performed on the average of the 80 target sweeps, from −1000 to +1000 ms with respect to the flash (visual stimulus), indicated on the abscissa as the time "0". Successively, the first and last 200 ms have been trimmed at both sides to remove edge effects. See the caption of the figure for explanations and the interpretation of results.

[1]The time of the day is also a circular quantity and provides a good example. The appropriate average of 22 h and 1 h is 23 h 30, but this is very far from their arithmetic mean. See also Cohen [14, pp. 214–246].

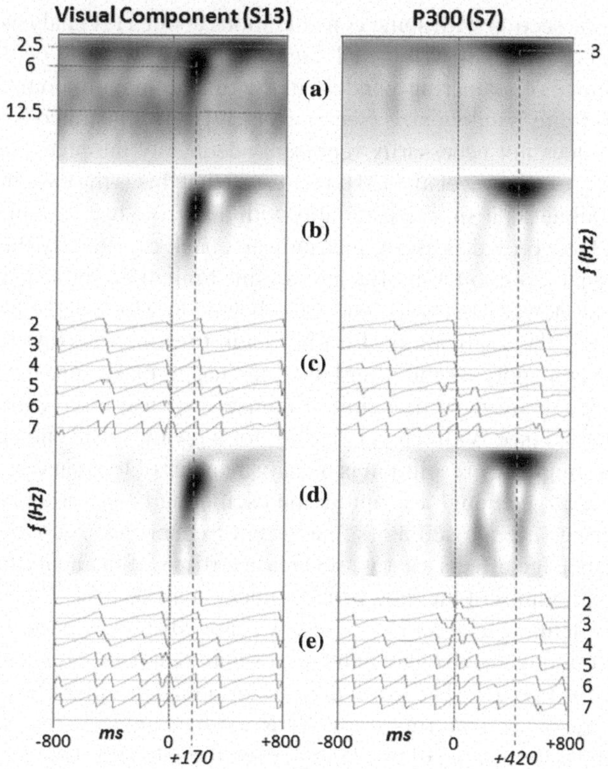

Fig. 4.7 Time-Frequency analysis of the source component S13 (left column) and S7 (right column) shown in Fig. 4.2. **a** Estimated instantaneous amplitude for frequency going from 1 Hz (top of the plot) to 20 Hz (bottom of the plot), in 0.5 Hz steps, computed using (4.17). This method is sensitive to *both phase-locked and non-phase-locked components*. The instantaneous amplitude is color coded, with white coding the minimum and black coding the maximum. The amplitude in **a** features a maximum in the time-frequency plane at around 6 Hz happening 170 ms post-stimulus, corresponding to the P100/N200 complex (see Fig. 4.2). We also notice a sustained activity around 2.5 Hz from about 200 to 700 ms post-stimulus. Note that at 2.5 Hz substantial power is present also *before* the stimulus, but this does not happen at 6 Hz. **b** Estimated instantaneous amplitude obtained with (4.15). This method is sensitive to *phase-locked components*. Note that both post-stimulus maxima at around 2.5 and 6 Hz survive, whereas anywhere else in the time-frequency plot the amplitude becomes negligible, including pre-stimulus activity around 2.5 Hz. Note also that the 2.5 Hz activity post-stimulus now is weaker. Taken together the analyses in **a** and **b** suggest that the activity around 6 Hz may be strictly phase-locked, whereas the activity at 2.5 Hz may be mixed with non-phase-locked components. Plot **c** shows the instantaneous phase of S13 in the closed interval $(-\pi \ldots -\pi]$, for frequencies in the range 2 Hz, ..., 7 Hz, in 1 Hz increments. This has been computed using (4.16), hence it is the phase spectrum corresponding to **b**. At about 220 ms post-stimulus, in correspondence to the end of the maximum at 6 Hz, the phase alines at all frequencies in the range 2 Hz, ..., 7 Hz. The amplitude spectrum in **d** and corresponding phase spectrum in **e** are the non-linear (normalized) version of **b** and **c**, respectively. The results are very similar to those seen in **b** and **c**, although they appear a bit noisier. For S7, the instantaneous amplitude (4.17) features only one strong maximum at about 3 Hz in between 280 and 570 ms (**a**, right column). This maximum corresponds to the P300 peak (Fig. 4.2). The same activity is seen also in **b** and **d**, although they appear noisier. This analysis suggests that the P300 is strictly phase-locked to the stimulus

We end up this section with some considerations about TFA analysis. The Hilbert transform can be obtained by the FFT algorithm [64]. The use of this algorithm requires the choice of a tapering window in the time domain to counteract spectral leakage due to finite window size (see Harris [39]). As illustrated in Fig. 4.4, the analytic signal does not necessarily represent adequately the phase of the original signal. The study of Chavez et al. [13] has stressed that this is the case in general only if the original signal is a simple oscillator with a narrow-band frequency support. These authors have provided useful measures to check empirically the goodness of the analytic signal representation. Because of this limitation, for a signal displaying multiple spectral power peaks or broad-band behavior, which is the case in general of EEG and ERP, the application of a filter bank to extract narrow-band behavior is necessary. When applying the filter bank, one has to make sure not to distort the phase of the signal. In general, a finite impulse response filter with linear phase response is adopted (see Widmann et al. [98], for a review). The choice of the filters band width and frequency resolution is usually a matter of trials and errors; the band width should be large enough to capture the oscillating behavior and small enough to avoid capturing several oscillators in adjacent frequencies. Also, the use of filter banks engenders edge effects, that is, severe distortions of the analytic signal at the left and right extremities of the time window under analysis [67]. This latter problem is easily solved defining a larger time window centered at the window of interest and successively trimming an adequate number of samples at both sizes, as we have done in the example of Fig. 4.7. The estimation of instantaneous phase for sweeps, time sample and frequencies featuring a low SNR are meaningless; the phase being an angle, it is defined for vectors of any length, even if the length (i.e., the amplitude) is negligible. However, phase measures can be interpreted only where the amplitude is high [7]. The effect is exacerbated if we apply the non-linear normalization, since in this case very small coefficients are weighted as the others in the average, whereas they should better be ignored.

4.5 Spatial Domain Analysis

Scalp topography and tomography (source localization) of ERPs are the basic tools to perform analysis in the *spatial domain* of the electrical activity generating ERPs. This is fundamental for linking experimental results to brain anatomy and physiology. It also represents an important dimension for studying ERP dynamics per se, complementing the information provided in time and/or frequency dimensions [52]. The spatial pattern of ERP scalp potential or of an ERP source component provides useful information to recognize and categorize ERP features, as well as to identify artifacts and background EEG. Early ERP research was carried out using only a few electrodes. Current research typically uses several tens and even hundreds of electrodes covering the whole scalp surface. More and more high-density EEG studies involve realistic head models for increasing the precision of source localization methods. Advanced spatial analysis has therefore become common practice in ERP research.

In contrast to continuous EEG, ERP studies allow spatial analysis with high-temporal resolution, i.e., they allow the generation of topographical and/or tomographical maps for each time sample. This is due to the SNR gain engendered by averaging across sweeps. Thus, as compared to continuous EEG, ERPs offer an analysis in the spatial domain with much higher temporal resolution. The SNR increases with the number of averaged sweeps. One can further increase the SNR by using a multivariate filtering method, as previously discussed. One can also increase the SNR by averaging spatial information in adjacent samples. The spatial patterns observed at all samples forming a peak in the *global field power*[2] can safely be averaged, since within the same peak the spatial pattern is supposed to be constant [52].

When using a source separation method (see Fig. 4.2 for an example) the spatial pattern related to each source component is given by the corresponding column vector of the estimated mixing matrix, i.e., the pseudo-inverse of the estimated matrix B^T. In fact, a source separation method decomposes the ensemble average in a number of source components, each one having a different and fixed spatial pattern. These patterns are analyzed separately as a topographic map and are fed individually to a source localization method as input data vector. Source localization methods in general perform well when the data is generated by one or two dipoles only, while if the data is generated by multiple dipoles the accuracy of the reconstruction is questionable [95]. BSS effectively decomposes the ensemble average in a number of simple source components, typically generated by one or two dipoles each [25]. As a consequence, spatial patterns decomposed by source separation can be localized with high accuracy by means of source localization methods. Note that applying a generic filtering method such as PCA and CSTP, the components given by the filter are still mixed and so are the spatial patterns held as column vectors by the matrix inverse of the spatial filter, that is, the pseudo-inverse of B^T. This prevents any physiological interpretation of the corresponding spatial patterns. Source separation methods are therefore optimal candidates for performing high-resolution spatial analysis by means of ERPs. An example of topographical analysis is presented in Figs. 4.2 and 4.8. For an example of tomographic analysis refer to Congedo et al. [24].

4.6 Inferential Statistics

As we have seen, in time-domain ERP studies it is of interest to localize experimental effects along the dimension of *space* (scalp location) and *time* (latency and duration of the ERP components). Analysis in the time-frequency-domain involves the study of amplitude and phase in the time-frequency plane. The dimensions retained by

---

[2]The global field power is defined for each time sample as the sum of the squares of the potential difference at all electrodes. It is very useful in ERP analysis to visualize ERP peaks regardless their spatial distribution [52].

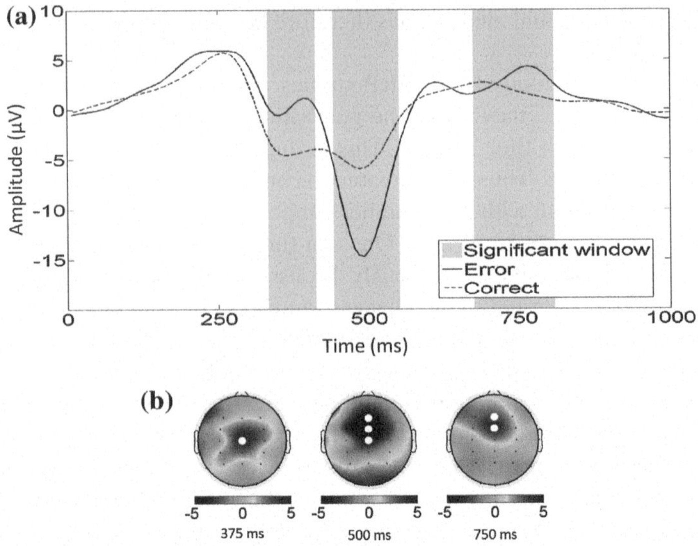

Fig. 4.8 **a** Ensemble average error-related potentials (19 subjects) for "Correct" and "Error" trials at electrode Cz. The supra-threshold cluster size permutation test was applied in the time and spatial dimension, with a = 0.05, to compare the "Correct" and "Error" condition. A significant positivity for error trials was found at time window 320–400 ms at electrode Cz ($p < 0.01$), a significant negativity for error trials at time window 450–550 ms at clustered electrodes Fz, FCz, Cz ($p < 0.01$) and a significant positivity for error trials at time 650–775 ms at clustered electrodes Fz, FCz ($p = 0.025$). Significant time windows are indicated by grey areas in (**a**) and significant clustered derivations by white disks in (**b**). The supra-threshold cluster size test display good power while controlling the FWER. Data is from the study of Congedo et al. [24]

the experimenter for the statistical analysis actually are combined to create a multi-dimensional measurement space. For example, if a time-frequency representation is chosen and amplitude is the variable of interest, the researcher defines a statistical hypothesis at the intersection of each time and frequency measurement point. Typical hypotheses in ERP studies concern differences in central location (mean or median) within and between subjects (t-tests), the generalization of these tests to multiple experimental factors including more than two levels, including their interaction (ANOVA) and the correlation between ERP variables and demographic or behavioral variables such as response-time, age of the participants, complexity of the cognitive task, etc. (linear and non-linear regression, ANCOVA).

The goal of a statistical test is to either reject or accept the corresponding *null hypothesis* for a given type I error (α), which is the a priori chosen probability to reject a null hypothesis when this is indeed true (false discovery). By definition, our conclusion will be wrong with probability α, which is typically set to 0.05. Things becomes more complicated when several tests are performed simultaneously; performing a statistical test independently for each hypothesis inflates the type I error rate proportionally to the number of tests. This is known as the *multiple-comparison problem* [41, 97] and is very common in ERP studies, where several points in time, space and frequency are to be investigated. Let M be the number of hypotheses to be tested and M_0 be the number of true null hypotheses. Testing each hypothesis independently at the α level, the expectation of false discoveries is $M_0 \times \alpha$. Thus, if all null hypotheses are actually true, i.e., $M_0 = M$, we expect to commit on the average $(100 \times \alpha)$ % false discoveries. This is, of course, an unacceptable error rate. Nonetheless, the more hypotheses are false and the more they are correlated, the more the error rate is reduced. ERP data is highly correlated along adjacent time points, spatial derivations and frequency. Therefore, special care should be undertaken in ERP statistical analysis to ensure that the error rate is controlled while preserving statistical power, that is, while preserving an acceptable chance to detect those null hypotheses that are false. Two families of statistical procedures have been employed in ERP studies with this aim: those controlling the *family-wise error rate (FWER)* and those controlling the *false-discovery rate (FDR)*.

The *family-wise error rate* (FWER) is the probability of making one or more false discoveries among all hypotheses. A procedure controlling the FWER at the α level ensures that the probability of committing even only one false discovery is less than or equal to α, regardless the number of tests and how many null hypotheses are actually true. The popular Bonferroni procedure belongs to this family; each hypothesis is tested at level α/M instead that at level α. Sequential Bonferroni-like procedures like the one proposed by Holm [42] also control the FWER, while featuring higher power. However, all Bonferroni-like procedures fail to take into consideration explicitly the correlation structure of the hypotheses, thus they are unduly conservative, the more so the higher the number of hypotheses to be tested.

An important general class of test procedures controlling the FWER is known as *p-min permutation tests* [75, 97], tracing back to the seminal work of Fisher [35] and Pitman [80–82]. Permutation tests are able to account adaptively for any correlation structure of hypotheses, regardless of its form and degree. Also, they do not need a distributional model for the observed variables, e.g., Gaussianity, as required by t-tests, ANOVA etc. [6, 30, 35, 43, 46, 75, 80–82, 91, 92, 97]. Even more appealing, one may extract whatever variable from the data and perform a valid test, thus we are not limited to test on central location, correlation, etc. Depending on the experimental design, even the random sampling assumption may be relaxed [30]. Given these characteristics, permutation tests are ideal options for testing hypotheses in ERP studies and have received much attention in the neuroimaging community [1, 43, 76].

Permutation tests are available for classical correlation, within- and between-subject mean difference tests, as well as for testing the main effects in ANOVA designs [30]. However, a straightforward permutation test for interaction effects in ANOVA designs does not exist, although some solutions have been proposed [75]. This is a major limitation if more than one independent variable is manipulated in the experiment. Also, like other resampling methods such as bootstrap and Monte Carlo, permutation tests require intense computations. For large data sets, permutation tests may be time consuming, although this is rarely a concern with modern computers and the typical size of data sets in ERP analysis.

Another kind of FWER-controlling permutation test in ERP analysis is the supra-threshold cluster size test [43]. A variant of this test has been implemented in the EEG toolbox Fieldtrip [71], following the review of Maris and Oostenveld [63]. This procedure assesses the probability to observe a concentration of the effect simultaneously along one or more studied dimensions. For example, in testing the mean amplitude difference of a P300 ERP, one expects the effect to be concentrated both along time, around 300–500 ms, and along space, at midline central and adjacent parietal locations. This leads to a typical correlation structure of hypothesis in ERP data; under the null hypothesis the effect would instead be scattered all over both time and spatial dimensions. An example of the supra-threshold cluster size test applied in the time-space ERP domain is shown in Fig. 4.8.

Another family of testing procedures controls the *false discovery rate* (FDR). The FDR is the expected proportion of falsely rejected hypotheses [3]. Indicating by R the number of rejected hypotheses and by F the number of those that have been falsely rejected, the FDR controls the expectation of the ratio F/R. This is clearly a less stringent criterion as compared to the FWER, since, as the number of discoveries increases, we allow proportionally more errors. The original FDR procedure of Benjamini and Hochberg [3] assumes that all hypotheses are independent, which is clearly not the case in general for ERP data. A later work has extended the FDR procedure to the case of arbitrary dependence structure among variables [4], however, contrary to what one would expect, the resulting procedure is more conservative, yielding low power in practice. The FDR procedure and its version for dependent hypotheses have been the subject of several improvements (e.g., [38, 87]). Recent research on FDR-controlling procedures attempts to increase their power by sorting the hypotheses based on a priori information [32]. Such sorting may be guided by previous findings in similar experiments, by the total variance of the variables when using central location tests, or by any criterion that is independent to the test-statistics. Another trend in this direction involves arranging the hypotheses in hierarchical trees prior to testing [102] and in analyzing experimental replicability [40]. The FDR procedures tend to be unduly conservative when the number of hypotheses is very large, although much less so than Bonferroni-like procedures. In contrast to FWER-controlling procedures, FDR-controlling procedures are much simpler and faster to compute. They offer, however, a much looser guarantee against the actual type I error rates and, like Bonferroni-like procedures, do not take explicitly into consideration the correlation structure of ERP data.

4.7 Single-Sweep Classification

The goal of a classification method is to automatically estimate the class to which a single-sweep belongs. The task is challenging because of the very low amplitude of ERPs as compared to the background EEG. Large artifacts, the non-stationary nature of EEG and inter-sweep variability exacerbate the difficulty of the task. Although single-sweep classification has been investigated since a long time [29], it has recently received a strong impulsion thanks to development of ERP-based brain computer interfaces (BCI: [101]). In fact, a popular family of such interfaces is based on the recognition of the P300 ERP. The most famous example is the *P300 Speller* [34], a system allowing the user to spell text without moving, but just by focusing attention on symbols (e.g., letters) that are flashed on a virtual keyboard.

The fundamental criterion for choosing a classification method is the achieved accuracy for the data at hand. However, other criteria may be relevant. In BCI systems, the training of the classifier starts with a calibration session carried out just before the actual session. Such calibration phase makes the usage of BCI system impractical and annoying. To avoid this, there are at least two other desirable characteristics that a classification method should possess [60]: its ability to *generalize* and its ability to *adapt*. Generalization allows the so-called *transfer learning*, thanks to which data from other sessions and/or other subjects can be used to initialize a BCI system so as to avoid the calibration phase. Transfer learning may involve using data from previous sessions of the same subject ("cross-session") and/or data from other subjects ("cross-subject"). The continuous (on-line) adaptation of the classifier [47, 48] ensures that optimal performance is achieved once the initialization is obtained by transfer learning [18]. Taken together, generalization and on-line adaptation ensure also the stability of the system in adverse situations, that is, when the SNR of the incoming data is low and when there are sudden environmental, instrumental or biological changes during the session. This is very important for effective use of a BCI outside the controlled environment of research laboratories.

Classification methods differ from each other in the way they define the *set of features* and in the *discriminant function* they employ. Traditionally, the classification approaches for ERPs have given emphasis to the optimization of either one or the other aspect in order to increase accuracy. The approaches emphasizing the definition of the set of features try to increase the SNR of single-sweeps by using multivariate filtering, as those we have encountered in the section on time domain analysis, but specifically designed to increase the separation of the classes in a reduced feature space where the filter projects the data [83, 104]. For data filtered in this way, the choice of the discriminant function is not critical, in the sense that similar accuracy is obtained using several types of discriminant functions. In general, these approaches perform well even if the training set is small, but generalize poorly across sessions and across subjects because the spatial filters are optimal only for the session and subject on whom they are estimated. Instead, the approaches emphasizing the discriminant function use sharp machine learning algorithms on raw data or on data that has underwent little-preprocessing. Many machine learning algorithms have been tried in the BCI literature for this purpose [59, 60]. The three traditional approaches that have been found effective in P300 single-sweep classification are the support-vector machine, the stepwise linear discriminant analysis and the Bayesian linear discriminant analysis. In general, those require large training sets and have high computational complexity, but generalize fairly well across sessions and across subjects. The use of a *random forest* classifier is currently gaining popularity in the BCI community, incited by good accuracy properties [33]. However, its generalization and adaptation capability have not been established yet. The *deep neural networks learning* [5] has recently been shown to be very promising in other fields of research. Studies testing its performance on ERP data are not conclusive so far. An approach that features at the same time good accuracy, good generalization and good adaptation capabilities in the case of ERP data has been recently borrowed from the field of *differential geometry*. This approach makes use of the *Riemannian geometry* on the manifold of symmetric positive definite (SPD) matrices. Covariance matrices are of this kind. A very simple classifier can be obtained based on the minimum distance to mean (MDM) method [2]: every sweep is represented as a covariance matrix, i.e., as a point on the multidimensional space of SPD matrices. The training set is used to estimate the *center of mass* of training points for each class, i.e., a point best representing the class. An unlabeled sweep is then simply assigned to the class the center of mass of which is the closest to the unlabeled sweep. This approach as well as other classifiers based on Riemannian geometry have been shown to possess good accuracy, generalization and robustness properties [19, 66, 103, 105]

References

1. S. Arndt, T. Cizadlo, N.C. Andreasen et al., Tests for comparing images based on randomization and permutation methods. J. Cereb. Blood Flow Metab. **16**, 1271–1279 (1996)
2. A. Barachant, S. Bonnet, M. Congedo et al., Multi-class brain computer interface classification by riemannian geometry. IEEE Trans. Biomed. Eng. **59**, 920–928 (2012)
3. Y. Benjamini, Y. Hochberg, Controlling the false discovery rate: a practical and powerful approach to multiple testing. J. Roy. Stat. Soc. Ser. B Stat. Methodol. **57**, 289–300 (1995)
4. Y. Benjamini, D. Yukutieli, The control of the false discovery rate in multiple testing under dependency. Ann. Stat. **29**, 1165–1188 (2001)
5. Y. Bengio, Learning deep architectures for AI. Found. Trends Mach. Learn. **2**, 1–12 (2009)
6. R.C. Blair, J.F. Troendle, R.W. Beck, Control of familywise errors in multiple assessments via stepwise permutation tests. Stat. Med. **15**, 1107–1121 (1996)
7. P. Bloomfield, *Fourier Analysis of Time Series. An Introduction*, 2nd edn. (Wiley, Hoboken, New Jersey, 2000), p. 261
8. A. Burns, Fourier-, Hilbert- and wavelet-based signal analysis: are they really independent approaches? J. Neurosci. Methods **137**, 321–332 (2004)
9. G. Buszáki, C.A. Anastassiou, C. Koch, The origin of extracellular fields and currents—EEG, ECoG, LFP and spikes. Nat. Rev. Neurosci. **13**, 407–420 (2012)
10. R.T. Canolty, E. Edwards, S.S. Dalal et al., High gamma power is phase-locked to theta oscillations in human neocortex. Science **313**, 1626–1628 (2006)
11. J.-F. Cardoso, Blind signal separation: statistical principles. Proc. IEEE **86**, 2009–2025 (1998)
12. R.M. Chapman, J.W. McCrary, EP component identification and measurement by principal component analysis. Brain Cogn. **27**, 288–310 (1995)
13. M. Chavez, M. Besserve, C. Adam et al., Towards a proper estimation of phase synchronization from time series. J. Neurosci. Methods **154**, 149–160 (2006)
14. M.X. Cohen, *Analyzing Neural Time Series Data: Theory and Practice* (The MIT Press, Cambridge, Massachusetts, 2014), p. 600
15. M.X. Cohen, J.F. Cavanagh, Single-trial regression elucidates the role of prefrontal theta oscillations in response conflict. Front. Psychol. **2**, 30 (2011)
16. L.L. Colgin, Theta-gamma coupling in the entorhinal-hippocampal system. Curr. Opin. Neurobiol. **31**, 45–50 (2015)
17. P. Comon, C. Jutten (eds.), *Handbook of Blind Source Separation, Independent Component Analysis and Applications* (Academic Press, Cambridge, MA, 2010)

18. M. Congedo, *EEG Source Analysis*. Dissertation, University of Grenoble Alpes, 2013
19. M. Congedo, A. Barachant, R. Bhatia, Riemannian geometry for EEG-based brain-computer interfaces; a primer and a review. BCI **4**, 155–174 (2017)
20. M. Congedo, C. Gouy-Pailler, C. Jutten, On the blind source separation of human electroencephalogram by approximate joint diagonalization of second order statistics. Clin. Neurophysiol. **119**, 2677–2686 (2008)
21. M. Congedo, M. Goyat, N. Tarrin et al., in *"Brain Invaders": A Prototype of an Open-Source P300- Based Video Game Working with the OpenViBE*. 5th International Brain-Computer Interface Conference, Graz, Austria, September 2011. (2011), pp. 280–283
22. M. Congedo, L. Korczowski, A. Delorme et al., Spatio-temporal common pattern; a reference companion method for ERP analysis. J. Neurosci. Methods **267**, 74–88 (2016)
23. M. Congedo, F.H. Lopes da Silva, Event-related potentials: general aspects of methodology and quantification, in *Niedermeyer's Electroencephalography, Basic Principles, Clinical Applications, and Related Fields*, ed. by D.L. Schomer, F.H. Lopes da Silva (Oxford University Press, Oxford, 2017)
24. M. Congedo, S. Rousseau, C. Jutten, An introduction to EEG source analysis with an illustration of a study on error-related potentials, in *Guide to Brain-Computer Music Interfacing*, ed. by E. Miranda, J. Castet (Springer, London, 2014), p. 313
25. A. Delorme, J. Palmer, J. Onton et al., Independent EEG sources are dipolar. PLoS One **7**, e30135 (2012)
26. A. Delorme, T. Sejnowski, S. Makeig, Enhanced detection of artifacts in EEG data using higher-order statistics and independent component analysis. Neuroimage **34**, 1443–1449 (2007)
27. J. Dien, Evaluating two-step PCA of ERP data with Geomin, Infomax, Oblimin, Promax, and Varimax rotations. Psychophysiology **47**, 170–183 (2010)
28. E. Donchin, A multivariate approach to the analysis of average evoked potentials. IEEE Trans. Biomed. Eng. **3**, 131–139 (1966)
29. E. Donchin, Discriminant analysis in average evoked response studies: the study of single trial data. Electroencephalogr. Clin. Neurophysiol. **27**, 311–314 (1969)
30. E.S. Edgington, *Randomization Tests*, 3rd edn. (Marcel Dekker, New York, 1995)
31. M. Falkenstein, J. Hohnsbein, J. Hoormann et al., Effects of crossmodal divided attention on late ERP components. II. Error processing in choice reaction tasks. Electroencephalogr. Clin. Neurophysiol. **78**, 447–455 (1991)
32. A. Farcomeni, L. Finos, FDR control with pseudo-gatekeeping based on possibly data driven order of The hypotheses. Biometrics **69**, 606–613 (2013)
33. F. Farooq, P. Kidmose, in *Random Forest Classification for P300 Based Brain Computer Interface Applications*. 21th European Signal Processing Conference, Marrakech, Morocco, September 2013. (2013) pp. 1–5
34. L.A. Farwell, E. Donchin, Talking off the top of your head: toward a mental prothesis utilizing event-related brain potentials. Electroencephalogr. Clin. Neurophysiol. **70**, 510–523 (1988)
35. R.A. Fisher, *Design of Experiments* (Oliver and Boyd, Edinburgh, 1935)
36. W.J. Freeman, Mechanism and significance of global coherence in scalp EEG. Curr. Opin. Neurobiol. **31**, 199–205 (2015)
37. D. Gabor, Theory of communication. J. IEE (London) **93**, 429–457 (1946)
38. W. Guo, M.B. Rao, On control of the false discovery rate under no assumption of dependency. J. Stat. Plan. Interference **138**, 3176–3188 (2008)
39. F.J. Harris, On the use of windows for harmonic analysis with the discrete Fourier transform. Proc. IEEE **66**, 51–83 (1978)
40. R. Heller, D. Yekutieli, Replicability analysis for genome-wide association studies. Ann. Appl. Stat. **8**, 481–498 (2014)
41. Y. Hochberg, A.C. Tamhane, *Multiple Comparison Procedures* (Wiley, Hoboken, NJ, 1987)
42. S. Holm, A simple sequentially rejective multiple test procedure. Scand. J. Stat. **6**, 65–70 (1979)

43. A.P. Holmes, R.C. Blair, J.D.G. Watson et al., Nonparametric analysis of statistic images from functional mapping experiments. J. Cereb. Blood Flow Metab. **16**, 7–22 (1996)
44. B.H. Jansen, G. Agarwal, A. Hedge et al., Phase synchronization of the ongoing EEG and auditory EP generation. Clin. Neurophysiol. **114**, 79–85 (2003)
45. E.R. John, D.S. Ruchkin, J. Vilegas, Experimental background: signal analysis and behavioral correlates of evoked potential configurations in cats. Ann. N. Y. Acad. Sci. **112**, 362–420 (1964)
46. W. Karniski, R.C. Blair, A.D. Snider, An exact statistical method for comparing topographic maps, with any number of subjects and electrodes. Brain Topogr. **6**, 203–210 (1994)
47. P.-J. Kindermans, D. Verstraeten, B. Schrauwen, A Bayesian Model for Exploiting Application Constraints to Enable Unsupervised Training of a P300-based BCI. PLoS ONE **7**, e33758 (2012)
48. P.-J. Kindermans, M. Schreuder, B. Schrauwen et al., True zero-training brain-computer interfacing—an online study. PLoS One **9**, e102504 (2014)
49. L. Korczowski, F. Bouchard, C. Jutten et al., in *Mining the Bilinear Structure of Data with Approximate Joint Diagonalization*. 24th European Signal Processing Conference, Budapest, Hungary, August 2016. (2016) pp. 667–671
50. J.-P. Lachaux, E. Rodriguez, J. Martinerie et al., Measuring phase synchrony in brain signals. Hum. Brain Mapp. **8**, 194–208 (1999)
51. T.D. Lagerlund, F.W. Sharbrough, N.E. Busacker, Spatial filtering of multichannel electroencephalographic recordings through principal component analysis by singular value decomposition. J. Clin. Neurophysiol. **14**, 73–82 (1997)
52. D. Lehmann, W. Skrandies, Reference-free identification of components of checkerboard-evoked multichannel potential fields. Electroencephalogr. Clin. Neurophysiol. **48**, 609–621 (1980)
53. M. Le Van Quyen, J. Foucher, J.-P. Lachaux et al., Comparison of Hilbert transform and wavelet methods for the analysis of neuronal synchrony. J. Neurosci. Methods **111**, 83–98 (2001)
54. J.E. Lisman, O. Jensen, The theta-gamma neural code. Neuron **77**, 1002–1016 (2013)
55. R.R. Llinas, The intrinsic electrophysiological properties of mammalian neurons: insights into central nervous system function. Science **242**, 1654–1664 (1988)
56. J.M. Lęski, Robust weighted averaging. IEEE Trans. Biomed. Eng. **49**, 796–804 (2002)
57. F.H. Lopes da Silva, Event-related neural activities: what about phase? Prog. Brain Res. **159**, 3–17 (2006)
58. F.H. Lopes da Silva, J.P. Pijn, D. Velis et al., Alpha rhythms: noise, dynamics, and models. Int. J. Psychophysiol. **26**, 237–249 (1997)
59. F. Lotte, M. Congedo, A. Lécuyer et al., A review of classification algorithms for EEG-based brain-computer interfaces. J. Neural Eng. **4**, R1–R13 (2007)
60. F. Lotte, L. Bougrain, A. Cichocki et al., A review of classification algorithms for EEG-based brain-computer interfaces: a 10-year update (Manuscript submitted for publication) (2018)
61. S. Luck, *An Introduction to the Event-Related Potential Technique*, 2nd edn. (The MIT Press, Cambridge, MA, 2014)
62. S. Makeig, M. Westerfield, T.P. Jung et al., Dynamic brain sources of visual evoked responses. Science **295**, 690–694 (2002)
63. E. Maris, R. Oostenveld, Nonparametric statistical testing of EEG- and MEG-data. J. Neurosci. Methods **164**, 177–190 (2007)
64. S.L. Marple, Computing the discrete-time analytic signal via FFT. IEEE Trans. Signal Process. **47**, 2600–2603 (1999)
65. A. Mazaheri, O. Jensen, Rhythmic pulsing: linking ongoing brain activity with evoked responses. Front. Hum. Neurosci. **4**, 177 (2010)
66. L. Mayaud, S. Cabanilles, A. Van Langhenhove et al., Brain-computer interface for the communication of acute patients: a feasibility study and a randomized controlled trial comparing performance with healthy participants and a traditional assistive device. BCI **3**, 197–215 (2016)

67. F. Mormann, K. Lehnertz, P. David et al., Mean phase coherence as a measure for phase synchronization and its application to the EEG of epilepsy patients. Physica D **144**, 358–369 (2000)
68. R. Näätänen, A.W.K. Gaillard, S. Mäntysalo, Early selective-attention effect on evoked potential reinterpreted. Acta Psychol (Amst) **42**, 313–329 (1978)
69. V.V. Nikulin, K. Linkenkaer-Hansen, G. Nolte et al., Non-zero mean and asymmetry of neuronal oscillations have different implications for evoked responses. Clin. Neurophysiol. **121**, 186–193 (2010)
70. P.L. Nunez, R. Srinivasan, *Electric Field of the Brain: The Neurophysics of EEG* (Oxford University Press, New York, 2006)
71. R. Oostenveld, P. Fries, E. Maris et al., FieldTrip: open source software for advanced analysis of MEG, EEG, and invasive electrophysiological data. Comput. Intell. Neurosci. **2011**, 1 (2011)
72. T.E. Özkurt, A. Schnitzler, A critical note on the definition of phase-amplitude cross-frequency coupling. J. Neurosci. Methods **201**, 438–443 (2011)
73. S. Palva, J.M. Palva, Discovering oscillatory interaction networks with M/EEG: challenges and breakthroughs. Trends Cogn Sci **16**, 219–230 (2012)
74. R.D. Pascual-Marqui, Instantaneous and lagged measurements of linear and nonlinear dependence between groups of multivariate time series: frequency decomposition. arXiv:0711.1455 (2007)
75. F. Pesarin, *Multivariate Permutation Tests* (Wiley, Hoboken, NJ, 2001)
76. K.M. Petersson, T.E. Nichols, J.-B. Poline et al., Statistical limitations in functional neuroimaging II. Signal detection and statistical inference. Philos. Trans. Roy. Soc. Lond. **354**, 1261–1281 (1999)
77. G. Pfurtscheller, F.H. Lopes da Silva, Event-related EEG/MEG synchronization and desynchronization: basic principles. Clin. Neurophysiol. **110**, 1842–1857 (1999)
78. D.-T. Pham, J.-F. Cardoso, Blind separation of instantaneous mixtures of non stationary sources. IEEE Trans. Signal Process. **49**, 1837–1848 (2001)
79. T.W. Picton, S. Bentin, P. Berg et al., Guidelines for using human event-related potentials to study cognition: recording standards and publication criteria. Psychophysiology **37**, 127–152 (2000)
80. E.J.G. Pitman, Significance tests which may be applied to samples from any population. J. Roy. Stat. Soc. Ser. B Stat. Methodol. **4**, 119–130 (1937)
81. E.J.G. Pitman, Significance tests which may be applied to samples from any population. II. The correlation coefficient. J. Roy. Stat. Soc. Ser. B Stat. Methodol. **4**, 225–232 (1937)
82. E.J.G. Pitman, Significance tests which may be applied to samples from any population. III. The analysis of Variance test. Biometrika **29**, 322–335 (1938)
83. B. Rivet, A. Souloumiac, V. Attina et al., xDAWN algorithm to enhance evoked potentials: application to brain-computer interface. IEEE Trans. Biomed. Eng. **56**, 2035–2043 (2009)
84. M.G. Rosenblum, A.S. Pikovsky, J. Kurths, Phase synchronization of chaotic oscillators. Phys. Rev. Lett. **76**, 1804–1807 (1996)
85. D.S. Ruchkin, An analysis of average response computatios based upon aperiodic stimuli. IEEE Trans. Biomed. Eng. **12**, 87–94 (1965)
86. S.C. Sereno, K. Rayner, Measuring word recognition in reading: eye-movements and event-related potentials. Trends Cogn. Sci. **7**, 489–493 (2003)
87. J.D. Storey, A direct approach to false discovery rate. J. Roy. Stat. Soc. Ser. B Stat. Methodol. **4**, 479–498 (2002)
88. S. Sutton, M. Braren, J. Zubin et al., Evoked-potential correlates of stimulus uncertainty. Science **150**, 1187–1188 (1965)
89. C. Tallon-Baudry, O. Bertrand, C. Delpuech et al., Stimulus specificity of phase-locked and non-Phase-locked 40 Hz visual responses in human. J. Neurosci. **16**, 4240–4249 (1996)
90. P. Tass, M.G. Rosenblum, J. Weule et al., Detection of n:m phase locking from noisy data: application to magnetoencephalography. Phys. Rev. Lett. **81**, 3291–3294 (1998)

91. J.F. Troendle, A stepwise resampling method of multiple hypothesis testing. J. Am. Stat. Assoc. **90**, 370–378 (1995)
92. J.F. Troendle, A permutation step-up method of testing multiple outcomes. Biometrics **952**, 846–859 (1996)
93. F. Varela, J.-P. Lachaux, E. Rodriguez et al., The brainweb: phase synchronization and large-scale integration. Nat. Rev. Neurosci. **2**, 229–239 (2001)
94. M. Vinck, R. Oostenveld, M. van Wingerden et al., An improved index of phase synchronization for electrophysiological data in the presence of volume-conduction, noise and sample-size bias. NeuroImage **55**, 1548–1565 (2011)
95. M. Wagner, M. Fuchs, J. Kastner, Evaluation of sLORETA in the presence of noise and multiple sources. Brain Topogr. **16**, 277–280 (2004)
96. W.G. Walter, R. Cooper, V.J. Aldridge et al., Contingent negative variation: an electric sign of sensorimotor association and expectancy in the human brain. Nature **203**, 380–384 (1964)
97. P.H. Westfall, S.S. Young, *Resampling-Based Multiple Testing. Examples and Methods for p-Values Adjustment* (Wiley, Hoboken, NJ, 1993)
98. A. Widmann, E. Schröger, B. Maess, Digital filter design for electrophysiological data—a practical approach. J. Neurosci. Methods **250**, 34–46 (2014)
99. M. Woldorff, Adjacent response overlap during the ERP averaging process and a technique (Adjar) for its estimation and removal. Psychophysiology **25**, 490 (1988)
100. M. Woldorff, Distortion of ERP averages due to overlap from temporally adjacent ERPs: analysis and correction. Psychophysiology **30**, 98–119 (1993)
101. J. Wolpaw, E.W. Wolpaw (eds.), *Brain-Computer Interfaces: Principles and Practice* (Oxford University Press, New York, 2012), p. 424
102. D. Yekutieli, Hierarchical false discovery rate-controlling methodology. J. Am. Stat. Assoc. **103**, 309–316 (2012)
103. F. Yger, M. Berar, F. Lotte, Riemannian approaches in brain-computer interfaces: a review. IEEE Trans. Neural Syst. Rehabil. Eng. **25**, 1753–1762 (2016)
104. K. Yu, K. Shen, S. Shao et al., Bilinear common spatial pattern for single-trial ERP-based rapid serial visual presentation triage. J. Neural Eng. **9**, 046013 (2012)
105. P. Zanini, M. Congedo, C. Jutten et al., Transfer learning: a riemannian geometry framework with applications to brain-computer interfaces. IEEE Trans. Biomed. Eng. (2017 in press)

Chapter 5
EEG Source Imaging and Multimodal Neuroimaging

Yingchun Zhang

Abstract During most common applications, the signals produced by cortical dipoles are detected at the scalp level. While these measurements can be highly informative and feature optimal temporal resolution, their spatial detection is hindered by the conduction through the tissues of the head. A deeper understanding of cortical sources can be provided by source imaging techniques. These methods first create mathematical models of the head, assigning appropriate properties to each layer. Brain activity is then calculated based on these assigned models and the observed EEG measurements, greatly improving the spatial resolution of EEG measurement and providing insights regarding otherwise hidden cortical dynamics. Source Imaging approaches can be further enhanced by integrating a second imaging modality. This is particularly useful with imaging methods that feature high spatial resolution or whose signals are not blurred by transduction. In the following chapter, we provide a detailed introduction to the general principles and basic algorithms of source imaging techniques. The discussion then expands to explore how other modalities can interact with these techniques to improve our results. At its conclusion, readers should have a good idea on how EEG data can be expanded to provide cortical insight.

After addressing the complexities of signal processing and analysis, the observation of EEG and ERP signals has been relatively straightforward. The conductive nature of these electrical signals opens up an intriguing new possibility for examining brain activity; using mathematical approaches to determine the most likely cortical sources of scalp potentials. Performing this backwards calculation is known as *electrical source imaging* (ESI). ESI techniques then present us with appealing imaging properties by allowing us to more directly observe brain activity while maintaining the low cost, high efficiency, and noninvasive features of EEG.

Before embarking on a detailed discussion of the algorithms and methods used in source imaging, it's first necessary to understand the cells and tissues within the

Y. Zhang (✉)
Department of Biomedical Engineering, University of Houston, Houston, USA
e-mail: yzhang94@uh.edu

C.-H. Im (ed.), *Computational EEG Analysis*, Biological and Medical Physics, Biomedical Engineering, https://doi.org/10.1007/978-981-13-0908-3_5

head. In previous chapters, we have addressed how EEG signals arise from the dipole moments of synaptic communication. We have also addressed how EEG signals are presented and detected. Calculating cortical source then requires us to mathematically determine the relationship between our synaptic signals and the recorded potentials. This will require us to give more attention to the biological environment of cortical activity, as well as the layers separating that source origin from the sensors on the scalp. By understanding these, we open the pathway to perform accurate electrical source imaging. Once we have this understanding, we can continue forward into the variety of methods that have been developed to observe brain activity and how EEG source analysis can interact with other imaging modalities to enhance our results. Finally, though the discussion below will focus on EEG-based source localization, it is important to acknowledge the process and principles described here are not exclusive and can be expanded for use with other modalities as well, such as electrocorticography (ECoG) [85] and electromyography (EMG) [50, 68, 86, 87].

5.1 From the Brain to the Scalp—The Forward Problem

When seeking to perform source analysis, the task can be broken down into two major parts: (i) the Forward Problem, which models the transference of putative source activity through the head to the scalp electrodes; and (ii) the Inverse Problem, which uses the information provided by the forward problem to identify the most likely locations and strengths of cortical activity. As mentioned above, we will begin with the cortical sources within the context of their conductive environment. From there, the discussion will build outward through the biological tissues to the scalp detection. At that point, we will finally be able to invert our process to observe the activity of unknown cortical sources.

5.1.1 Volume Source and the Poisson's Equation

Let us first examine an active synapse, wherein a pre-synaptic axon is communicating with a post-synaptic dendrite. The small volume enclosing this synapse can be assigned an overall current density J. The electrical model for the neuronal activity can then be described using two current monopoles: (i) a current source at the axon of a cell that injects positive ions into the extracellular space and (ii) a current sink at the coinciding dendrite that removes positive ions from the extracellular space [38]. Over time, the net current entering and leaving this volume must be zero to ensure that charge does not amass in the extracellular space, thus, $\nabla \cdot J = 0$.

This current density J comprises the *primary* current J_p (also known as the *impressed* current) [54] and the *volume* current J_v (also known as the *return* current) [54]. The primary current is generated by the movement of ions across the neuronal membrane, while the volume current is the Ohmic return current caused from the primary current that completes the circuit.

$$J = J_p + J_v \tag{5.1}$$

The electric field resulting from the volume current is given by Ohm's law as follows:

$$J_v = \sigma E \tag{5.2}$$

And the relationship between the electric field and potential field is given, under quasi-static conditions ($\nabla \times E = 0$), is given as:

$$E = -\nabla(V) \tag{5.3}$$

Thus, from (5.1), (5.2), and (5.3), the current density generated by neuronal activity is:

$$J = J_p - \sigma \nabla(V) \tag{5.4}$$

Taking the divergence of (5.4), we obtain:

$$\nabla \cdot J = \nabla \cdot J_p - \nabla \cdot (\sigma \nabla(V)) \tag{5.5}$$

Considering that $\nabla \cdot J = 0$, (5.5) becomes the Poisson's equation:

$$\nabla \cdot J_p = \nabla \cdot (\sigma \nabla(V)) \tag{5.6}$$

With the solution for this equation derived as [54]:

$$V = \frac{1}{4\pi\sigma} \int_v J_p \cdot \nabla\left(\frac{1}{r}\right) dv \tag{5.7}$$

As the current propagates throughout different brain tissue compartments, two characteristic behaviors of electrical charge at tissue boundaries are established: (i) all charge leaving one compartment (with a conductivity of σ_1) must enter the second compartment (assigned a conductivity value of σ_2):

$$\sigma_1 \nabla(V_1)\vec{n} = \sigma_2 \nabla(V_2)\vec{n} \tag{5.8}$$

where $\vec{n}$ is the normal vector at the boundary of the interface; and (ii) no current leaves the outer surface interfacing with air:

$$\sigma_{\text{outer}} \nabla(V_{outer})\vec{n} = 0 \tag{5.9}$$

Equations (5.8) and (5.9) are referred to as the Neumann boundary conditions. Now that we have established some of the necessary parameters ruling tissue conductance in the scalp, we can return to a more in-depth discussion of our cortical dipole itself.

5.1.2 *The Current Dipole*

The primary current, J_p, can be used to describe the current flow characteristics of any large group of pyramidal cells in a small cortical region that gives rise to EEG measurements. The source of this primary current can be further modeled as a single *equivalent current dipole* with two monopoles located at r_{source} and r_{sink} that are given opposite signs but an equal strength of I. The dipole position r_{dip} can therefore be described as the midpoint between r_{source} and r_{sink}. The dipole moment d of a dipole with the current I and an inter-pole distance of l is defined by an orientation unit vector e_d with the magnitude $\|d\| = I \cdot l$. This dipole moment can be further decomposed into $d = d_x e_x + d_y e_y + d_z e_z$, where e_x, e_y, e_z are unit vectors along the Cartesian axes and d_x, d_y, d_z are magnitudes of the respective dipole moment components. Thus, a single current dipole consists of 6 parameters: r_{dip}—which includes the 3 position parameters—and d—which accounts for the 3 dipole moment parameters. A potential field at position r generated by the current dipole d at position r_{dip} in an infinite, homogeneous volume conductor with a conductivity of σ is then calculated using the following equation:

$$V(r, r_{dip}, d) = \frac{d \cdot (r - r_{dip})}{4\pi\sigma \|r - r_{dip}\|^3} \tag{5.10}$$

5.1.3 *The Forward Problem—Algebra*

Generally speaking, the EEG forward problem aims to formulate the potential at any arbitrary scalp position that can be generated by any current dipole (r_{dip}, d) in the brain:

$$V(r) = g(r, r_{dip}, d) = g(r, r_{dip}, e_d) \cdot \|d\| \tag{5.11}$$

The function $g(r, r_{dip}, e_d)$ describing the measured scalp voltage at an arbitrary point r, which is generated by a current dipole with position r_{dip} and moment d, is formulated by solving the aforementioned Poisson's equation. Following the principle of superposition, a scalp potential generated by multiple current dipole sources according to (5.11) becomes:

$$V(r) = \sum_i g\left(r_i, r_{dip_i}, e_{d_i}\right)\|d_i\| \tag{5.12}$$

For N number of scalp measurements and P number of current dipoles over T number of discrete time samples, (5.12) is written in vector form as:

$$\begin{bmatrix} V(r_1, 1) & \cdots & V(r_1, T) \\ \vdots & \ddots & \vdots \\ V(r_N, 1) & \cdots & V(r_N, T) \end{bmatrix} = \begin{bmatrix} g\left(r_1, r_{dip_1}, e_{d_1}\right) & \cdots & g\left(r_1, r_{dip_P}, e_{d_P}\right) \\ \vdots & \ddots & \vdots \\ g\left(r_N, r_{dip_1}, e_{d_1}\right) & \cdots & g\left(r_N, r_{dip_P}, e_{d_P}\right) \end{bmatrix} \begin{bmatrix} \|d_{1,1}\| & \cdots & \|d_{1,T}\| \\ \vdots & \ddots & \vdots \\ \|d_{P,1}\| & \cdots & \|d_{P,T}\| \end{bmatrix} \tag{5.13}$$

or in matrix form as:

$$\boldsymbol{V} = \boldsymbol{G}\boldsymbol{J} \tag{5.14}$$

A noise or perturbation matrix $\mathbf{n}$ is then added to (5.15) to formulate what is typically termed the forward equation:

$$\boldsymbol{V} = \boldsymbol{G}\boldsymbol{J} + \boldsymbol{n} \tag{5.15}$$

where $\boldsymbol{V} \in \mathbb{R}^{N\times T}$ is the measurement matrix with N electrodes and T time samples, $\boldsymbol{G} \in \mathbb{R}^{N\times P}$ is the *gain* or the *lead-field matrix,* and $\boldsymbol{J} \in \mathbb{R}^{P\times T}$ depicts the dipole magnitude of P dipoles over T time samples. Solving the EEG forward problem amounts to the computation of the coefficients of the $\boldsymbol{G}$ matrix, given the locations and configurations of the dipole sources, recording electrodes, and the characteristics of the volume conductor.

5.1.4 The Volume Conductor: Type of Models

In the conventional approach, the transfer-coefficients making up the matrix G in (5.15) are obtained by calculating the surface potentials from dipole sources via Poisson's equation. These calculations are made for each dipole position within the head model and the resulting potentials are recorded at the electrode positions.

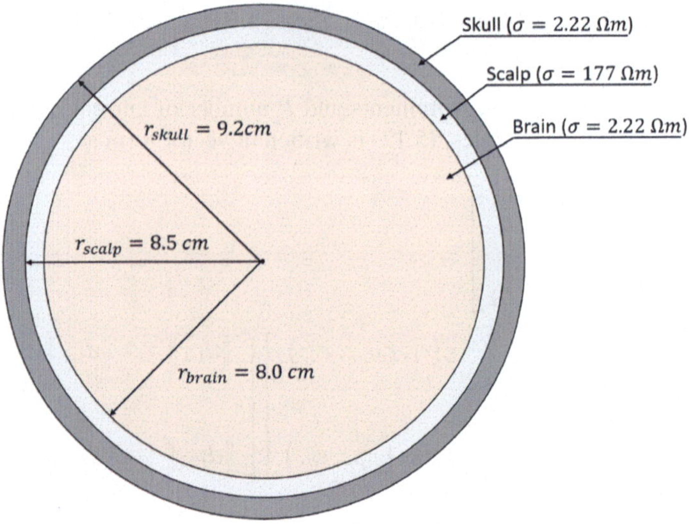

Fig. 5.1 A three-layer concentric spherical model of the volume conductor. Each layer represent a head compartment: brain, skull, scalp with the corresponding conductivity values [73]

5.1.4.1 Spherical Head Model

In its most simple construction, the head can be modeled as a single-layer homogenous, isotropic, conductive sphere. This greatly simplifies the computation of source localization, as the relationship between observed scalp potentials and cortical dipole(s) can be solved directly using (5.10). However, it is not hard to see that this is an oversimplified head model—human heads are neither spherical nor homogenous. Ignoring realistic geometries for now, the head model can be improved by establishing three concentric, spherical regions representing the brain, skull, and scalp and assigning each an appropriate conductivity value (see [73] and Fig. 5.1). Partly due to the simplicity of the spherical shape, there exists an analytical solution for this three-layer concentric spherical model, derived from the Poisson's equation (see [74]). Thus, the analytical solution of this spherical model is extensively used to test and validate the performance of more sophisticated numerical methods on complex volume conductor models (described in the following sections).

While addressing the inhomogeneity of the head improves the cortical model, the achieved solution from a spherical head model will still fail to accurately reconstruct signals until the geometrical complexity of the head (and its constituent tissues) is addressed. This is particularly important, as the thickness of tissue layers may vary and the complex curvatures of the gyri and sulci of the brain can significantly affect the solution. To fully capture the geometrical aspects of the brain when modelling the volume conductor, the use of realistic head models is becoming common practice in the field of EEG forward modeling.

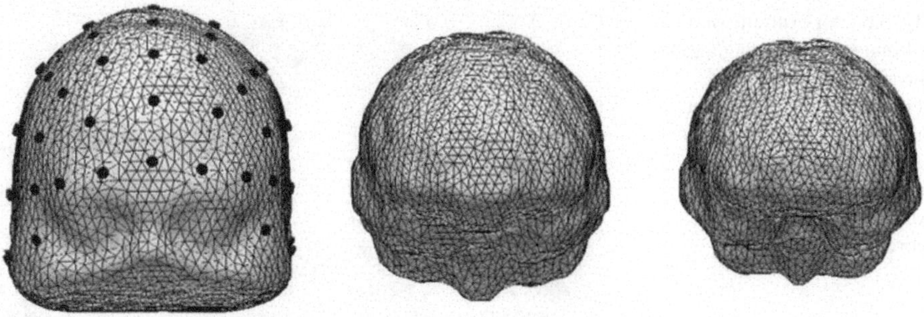

Fig. 5.2 The surface model from a realistic geometry human head. The three layers, from left to right, are the scalp, skull, and brain. Figure reproduced from [30]

5.1.4.2 Realistic-Head Model

Typically, the volume conductor constructed from a realistic-head model is comprised of the same 3 surfaces discussed above: the brain, skull, and skin surface. The overall process starts at the acquisition of images from structural scans of the brain, such as CT or MRI images. These images are segmented into multiple surfaces in the form of closed triangular meshes with a finite number of nodes, preserving the structural integrity of the actual head. The brain, skull, and skin surfaces serve as boundary layers that encapsulate the volumes of specific tissues (see Fig. 5.2). Homogeneous volume conductive properties can then be assigned to each defined tissue volume.

Given the geometric complexity of this head model, the potential at any node on the scalp surface generated by the dipoles in the brain compartment can be estimated using a numerical technique called the *boundary element method* (BEM). While the detailed formulation of the BEM can be found elsewhere [38, 57], we will consider the resultant primary equation that describes the potential distribution at each boundary surface:

$$v = g + \boldsymbol{B}v \tag{5.16}$$

where v^i is the potential value at the ith vertex, g^i is the potential due to the source at the ith vertex, and matrix $\boldsymbol{B}$ represents the dependency of each vertex point on each other based on the geometry of the surfaces and conductivities of each compartment. The accuracy of this estimation partly depends on the resolution of the surfaces (i.e. the number of nodes and triangle meshes that make up the three compartment surfaces) [30].

The BEM modelling of the human head volume conductor is inherently limited in its capability to capture the anisotropic properties and local inhomogeneities of biological brain tissues. This is due to the nature of the BEM calculation—changes in tissue properties and their effects on conducted signals are only implemented at the surface interfaces, and constant conductivity values are assumed for the spaces in between these boundaries. In cases where the human head volume conductor needs

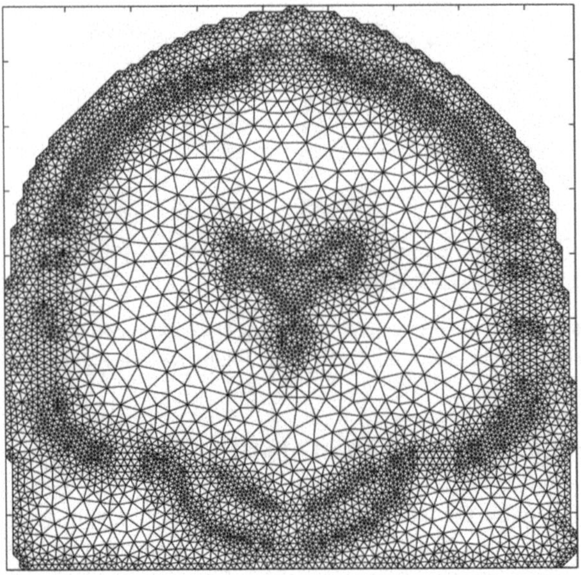

Fig. 5.3 An example of a volume mesh showing a 2D coronal slice. This 3D digitization of the head forms the basis for finite-element method for the 3D volume model of the human head volume conductor. Figure reproduced from [38]

to be represented as a proper volume (particularly, the presence of altered tissue properties or implanted devices [83, 84]), the solution to the forward model in a 3D volume can instead be obtained through the use of the finite element method (FEM) [81, 82]. Figure 5.3 shows a 2D coronal slice digitization of the head used in FEM.

Following the FEM, the potential V is calculated at each node of the 3D mesh as:

$$V(x, y, z) = \sum_{i=1}^{n} V_i \varphi_i(x, y, z) \tag{5.17}$$

where $\varphi_i(x, y, z)$ is a set of basis functions, n is the total number of vertices in the entire volume conductor Ω, and V_i is the potential associated with the ith node. Along with the Neumann boundary condition (5.9), the "weak formulation" for the FEM is obtained based on the Galerkin approach as:

$$-\int_{\Omega} \nabla\varphi \cdot (\sigma \nabla(V)) d\Omega = \int_{\Omega} \nabla \cdot J_p d\Omega \tag{5.18}$$

This formulation can be written as a linear system of equations in matrix form by substituting (5.17) into (5.18) (see the detailed derivations in [56, 81, 82]):

$$\boldsymbol{K}\boldsymbol{V} = \underline{\mathbf{J}} \tag{5.19}$$

where $\boldsymbol{K}$ refers to the stiffness matrix, incorporating the geometry and conductivity properties of the volume conductor, $\boldsymbol{V}$ is the potential vector at each of the nodes in Ω, and $\underline{\mathbf{J}}$ is the source term obtained from the right hand side integration of (5.18). This system of equations can then be solved using a number of iterative linear methods. A comprehensive review of these methods is given in [38].

5.1.5 A Note on Tissue Conductivity Values

The conductivity values for the different tissues in the brain contribute significantly to the accuracy of the forward model. The conductivity for the cerebro-spinal fluid (CSF) is widely accepted to be 1.79 S/m [8]. The conductivity values for the brain and scalp compartment were also reported to be 0.33 S/m with good agreement in the field [33, 36, 45], while the soft tissue to skull conductivity ratio remains a subject of debate, leading to a multitude of studies conducted to arrive at a more accurate number. For instance, the brain-to-skull ratio was originally suggested to be 80 by multiple research groups using different analysis techniques [16, 31, 72]. More recently, however, the ratio was estimated to be 15 based on in vitro and in vivo experiments performed by Oostendorp et al. [62]. In 2004, Guttierrez et al. reported the soft tissue to skull conductivity of 26 using EEG scalp measurement and a 4-sphere head model [36], while Lai et al. suggested using spherical head model a ratio of 25 from the in vivo cortical imaging of 5 epilepsy patients in 2005 [45]. Subsequently, using a realistic geometry inhomogeneous head model, Zhang et al. estimated a brain-to-skull conductivity ratio of 18.7 through in vivo experiments of intracranial electrical stimulation in two epilepsy patients [83, 84].

5.2 From the Scalp to the Brain—The Inverse Problem

Now that we have defined the relationship between dipoles and the potentials they generate on the scalp layer, we can attempt to invert this relationship to determine which parts of the brain are active from their associated scalp potentials. This process is referred to as EEG source localization. EEG source localization begins at the forward problem (see 5.15):

$$\boldsymbol{V} = \boldsymbol{G}\boldsymbol{J} + \boldsymbol{n}$$

and formulates the inverse problem as:

$$\hat{\boldsymbol{J}} = \boldsymbol{M}\boldsymbol{V} \tag{5.20}$$

where $\boldsymbol{V} \in \mathbb{R}^{N \times T}$ is the measurement matrix of N electrodes and T time samples, $\boldsymbol{G} \in \mathbb{R}^{N \times P}$ is the gain or the lead-field matrix, $\boldsymbol{J} \in \mathbb{R}^{P \times T}$ depicts the dipole magnitude of P dipoles over T time samples, $\boldsymbol{M} \in \mathbb{R}^{P \times N}$ is the inverse operator, and $\hat{\boldsymbol{J}}$ is the estimated $\boldsymbol{J}$. As such, the EEG inverse problem amounts to the calculation of matrix $\boldsymbol{M}$ that would result in a satisfactory value for $\hat{\boldsymbol{J}}$ and the best fit with the observed scalp potentials. The EEG inverse problem is an ill-posed problem due to the non-uniqueness of its solution (the number of unknown sources is much larger than the number of scalp measurements; $P \gg N$). To this end, there exists a number of mathematical optimization schemes that provide solutions to the inverse problem depending on the different configurations of the forward model. Specifically, different assumptions can be made regarding the properties of the source space, such as the number and/or the spatial distribution of the source current dipoles, and whether the positions, magnitudes, and orientations of potential dipoles are fixed or varied. Given the different forward model configurations, there are two main approaches to the EEG inverse problem: parametric and non-parametric optimization methods.

5.2.1 Parametric Optimization Methods

In the family of parametric optimization methods, the source space usually comprises a single dipole or a few dipoles with unknown position(s), magnitude(s), and orientation(s). These configurations are also known as *equivalent current dipole* (ECD) models, wherein the solutions are obtained by searching for the "equivalent dipole(s)" that best explain the observed scalp potentials. We present in the following sections several popular representative methods for the EEG inverse problem of the ECD model.

5.2.1.1 The Least-Squares Method

In a source model involving a single dipole, the solution can be obtained using the non-linear least-squares method. This method solves for a single dipole with an unknown position and moment that result in a global minimization of the residual error between the estimated and observed EEG signals. Thus, the cost function involving the position and moment parameters can be written in the form of an L_2-norm as follows:

$$\min_{(r_{dip},d)} \|\boldsymbol{V} - \left(\boldsymbol{G}\hat{\boldsymbol{J}} + \boldsymbol{n}\right)\|_2^2 \tag{5.21}$$

An iterative process is then employed to search for the best fit solution. This process begins by fixing the dipole at an estimated position, then adjusting the dipole orientation and magnitude and computing the least-squares error. A new dipole source position is then selected and the process is repeated until a global minimum least-square error is achieved, yielding the solution dipole with the best-fit position and moment. This non-linear least-squares method can be extended for multiple dipoles in a similar fashion, where the number of potential dipoles is chosen by the operator before the iterative process. With each iterative step requiring multiple calculations of the forward solution using the test dipole(s), the least-squares method has a high computational demand that is greatly increased when attempting to model multiple dipoles. To assuage this, several search methods have been commonly implemented—including the gradient, Nelder-Meade downhill search, multi-start simplex, genetic algorithms [78]. Furthermore, the least-squares method faces two significant drawbacks: (i) the solution with the minimum least-squares error is not necessarily closest to the true underlying sources and (ii) The true number of active dipoles is unknown and likely too large to be represented by a single "equivalent dipole".

5.2.1.2 Beamforming Approaches

Another class of the parametric inverse solution is the beamforming methods. Unlike the least-squares method, beamforming approaches do not search for equivalent dipole(s) that fully explain the measured potentials. Instead, the contribution of a single dipole source to the detected field is estimated, meaning that the number of dipoles does not have to be assumed a priori. The beamformer acts as a spatial filter that monitors the activity originating from one dipole source of interest and filters out contributions from all other sources. The 3-element vector of the dipole moment (the x-, y-, and z-components of the dipole moment) representing the contribution of dipole source at the known position r_{dip} is estimated using the following formulation:

$$y = \boldsymbol{W}^T(r_{dip})v(t) \tag{5.22}$$

where y is the 3-component dipole moment vector, v is the scalp potential measurements at time t, and $\boldsymbol{W}$ is the spatial filter matrix for dipole source at position r_{dip}. Ideally, the spatial filter is designed as a passband that selects only the sources within a small distance δ from r_{dip} while serving as a stopband for sources elsewhere, thus it must satisfy the following constraints on the forward model $\boldsymbol{G}$:

$$\boldsymbol{W}^T(r_{dip})\boldsymbol{G}(r) = \begin{cases} \boldsymbol{I} & for \|r_{dip} - r\| \le \delta \\ 0 & for \|r_{dip} - r\| > \delta \end{cases} \tag{5.23}$$

Using the linearly constrained minimum variance (LCMV) approach [80], the estimation of $\boldsymbol{W}$ amounts to:

$$\min_{\boldsymbol{W}^T} tr\{C_y\} \quad subject\ to \quad \boldsymbol{W}^T(r_{dip})\boldsymbol{G}(r_{dip}) = \boldsymbol{I} \tag{5.24}$$

where $C_y = \boldsymbol{W}^T C_v \boldsymbol{W}$, and C_v is the signal covariance matrix obtained from the measurements, and tr denotes the trace of a matrix. Applying the Lagrange multipliers, the solution for $\boldsymbol{W}$ can be derived as [79]:

$$\boldsymbol{W}(r_{dip}) = \left[\boldsymbol{G}(r_{dip})^T C_v^{-1}\boldsymbol{G}(r_{dip})\right]^{-1}\boldsymbol{G}(r_{dip})^T C_v^{-1} \tag{5.25}$$

To obtain the time-course of activity of the current source at r_{dip}, one can apply the spatial filter matrix $\boldsymbol{W}$ to each of the measurement vectors $v(t)$ for all $t = 1, \ldots, T$. Furthermore, this approach can reconstruct the dipole activity at any location by simply changing the r_{dip}, so long as the new source position is anatomically realistic. Beamforming techniques may, however, struggle in cases where highly correlated, spatially distinct sources are active [70, 80]. A straightforward strategy to handle this problem is the introduction of longer time windows, as the high correlation between sources is less likely to persist across the longer period. Brookes et al. proposed a modified source model in a *dual source beamformer* technique that could successfully reconstruct correlated sources in a simulation study [12], while other simulation studies have suggested that the LCMV method is robust against moderate level of source correlation [76, 80].

5.2.2 Non-parametric Optimization Methods

In contrast to ECD models, which assume that underlying sources can be represented by a single or small group of equivalent dipoles that explain the observed measurements, non-parametric optimization methods make use of *cortically distributed source* (CDS) models. The CDS models base on the assumption that the primary current sources are the cortical pyramidal neurons that span the cortex and orient normally to the surface. Thus, the source space is constructed with a current dipole assigned at each of the mesh element of the cortical layer, with dipole orientations either fixed to the local surface normal or left as unknown. In practice, cortical surface models are usually extracted from brain anatomical images (e.g. MRI, CT) through segmentation algorithms [17, 18, 24, 25]. In this setting, the dipole locations are known and the parameter of interest is the dipole moment of each source location. The number of source points on the cortical mesh will vary depending on the chosen model and mesh element properties, but is typically on the order of several thousands. This again makes the inverse problem highly underdetermined ($P \gg N$). The inverse solutions are thus obtained by applying some form of regularization to the cost function. The formulation of the EEG inverse problem is presented again below, with the various regularization schemes and their limitations described in subsequent sections. Consider the following linear formulations of the forward and inverse problems:

$$\boldsymbol{V} = \boldsymbol{G}\boldsymbol{J} + \boldsymbol{n}$$

$$\hat{\boldsymbol{J}} = \boldsymbol{M}\boldsymbol{V}$$

where we assume a priori that statistical distributions of the dipole moment $\boldsymbol{J}$ and sensor vector $\boldsymbol{n}$ exist such that $\boldsymbol{J} \sim \mathcal{N}(0, \boldsymbol{R})$ and $\boldsymbol{n} \sim \mathcal{N}(0, \boldsymbol{C})$. Matrix $\boldsymbol{M}$ is the linear inverse operator that maps the EEG measurements $\boldsymbol{V}$ into the estimated source $\hat{\boldsymbol{J}}$.

The matrix multiplication of the forward and inverse operator, $\boldsymbol{GM}$, is referred to as the resolution matrix and ideally equals to the identity matrix $\boldsymbol{I}$, representing a perfect inverse solution. The inverse problem then amounts to the following generalized cost function:

$$\min_{\boldsymbol{M}}\left\{\|\left(\boldsymbol{G}\hat{\boldsymbol{J}}+\boldsymbol{n}\right)-\boldsymbol{V}\|_p^2+\lambda \boldsymbol{L}\left(\hat{\boldsymbol{J}}\right)\right\} \tag{5.26}$$

where $\|\cdot\|_p$ denotes the L_p-norm and $\boldsymbol{L}\left(\hat{\boldsymbol{J}}\right)$ is some regularization scheme or a priori constraint on the estimated sources with a scalar regularization parameter λ. The regularization parameter λ represents the balance between maximizing the goodness of fit and minimizing the constraint term $\boldsymbol{L}\left(\hat{\boldsymbol{J}}\right)$. λ is typically selected using, among others [35], two common approaches: the generalized cross-validation (GCV) method [32] and the L-curve method [40]. The optimal value for λ is obtained in the GCV method by minimizing a function of λ:

$$GCV(\lambda)=\frac{\|\left(\boldsymbol{G}\hat{\boldsymbol{J}}(\lambda)+\boldsymbol{n}\right)-\boldsymbol{V}\|^2}{(tr(\boldsymbol{I}-\boldsymbol{GM}))^2} \tag{5.27}$$

The numerator represents the residual error resulting from a solution $\hat{\boldsymbol{J}}$ regularized by a particular λ value, while the denominator depicts the inaccuracy in the resolution matrix. On the other hand, the L-curve is a log-plot of the norm of the residual error term $\left(\boldsymbol{G}\hat{\boldsymbol{J}}+\boldsymbol{n}\right)-\boldsymbol{V}$ against the norm of the regularized solution term $\boldsymbol{L}\left(\hat{\boldsymbol{J}}\right)$ at multiple values for the regularization parameter λ. Figure 5.4 demonstrates the L-curve plot and the effect of different λ values on the inverse solution. Typically, a regularization parameter λ is chosen to be near the "characteristic corner" of the L-curve and, as such, generally yields a good balance between a small residual norm and a small solution norm.

5.2.2.1 The Minimum-Norm Estimates

The *minimum-norm estimates* (MNE) [39] produces the inverse solution that minimize the overall power of the estimated source activity. Here, the L_2-norm is applied on the error term, and $\boldsymbol{L}\left(\hat{\boldsymbol{J}}\right)=\|\hat{\boldsymbol{J}}\|_2^2$, yielding the solutions as:

$$\boldsymbol{M}_{MNE}=\boldsymbol{G}^T\left(\boldsymbol{G}\boldsymbol{G}^T+\lambda\boldsymbol{I}\right)^{-1} \tag{5.28}$$

In such expression, the source and noise covariance matrices, $\boldsymbol{R}$, and $\boldsymbol{C}$, are assumed to be an identity matrix $\boldsymbol{I}$. A more generalized expression of MNE that explicitly accounts for the covariance matrix is given as [17]:

$$\boldsymbol{M}_{MNE}=\boldsymbol{R}\boldsymbol{G}^T\left(\boldsymbol{G}\boldsymbol{R}\boldsymbol{G}^T+\lambda\boldsymbol{C}\right)^{-1} \tag{5.29}$$

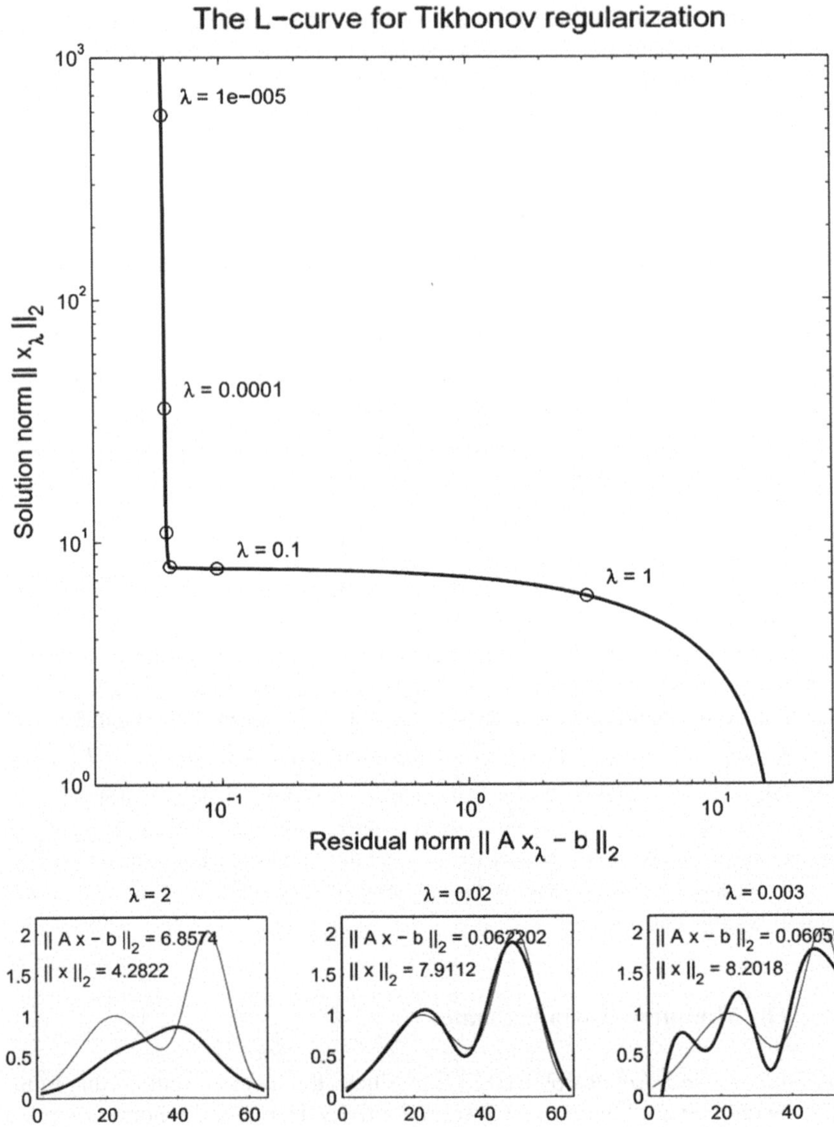

Fig. 5.4 The L-curve method for selecting the optimal value for the regularization parameter λ. Top—the plot of the norm of the regularized solution norm against the norm of the residual at various value of λ. Bottom—the fitting of the inverse solution (thicker-line) to the ground-truth (thinner-line) at under-constrained (left), optimally constrainted (middle) and over-constrained (right) values of λ. Figure reproduced from [40]

This general expression for the MNE solution can be shown to be equivalent to the Tikhonov regularization method, as well as the solution derived using a Bayesian framework (for a detailed derivation and description, see [35]). The MNE solution is generally suitable for the estimation of superficially distributed sources; however, MNE is also known to underestimate deeper sources.

To alleviate the depth-bias of MNE, a *weighted minimum-norm estimates* (wMNE) algorithm was introduced, in which:

$$L\left(\hat{J}\right) = \|W\hat{J}\|_2^2 \tag{5.30}$$

where W is the weighting matrix used to compensate for deeper sources, constructed from the norm of each column of the lead field matrix G. This is expressed as:

$$W = \Omega \otimes \mathbf{I}_3 \tag{5.31}$$

where Ω is a diagonal matrix of size $P \times P$, Ω_{ii} equals to the norm of the ith column of the lead field matrix, and the inverse operator is expressed as:

$$M_{wMNE} = \left(W^T W\right)^{-1} G^T \left(G\left(W^T W\right)^{-1} G^T + \lambda I\right)^{-1} \tag{5.32}$$

5.2.2.2 Low Resolution Electromagnetic Tomography

The *low resolution electromagnetic tomography* (LORETA) method [64] aims to incorporate physiological principles into the EEG inverse solution; specifically, the idea that neighboring neural sources are activated simultaneously and synchronously. To this end, LORETA implements a smoothing operation on the source space in the form of a Laplacian operator. The inverse solution obtained from the LORETA method is then presented as:

$$\begin{gathered} M_{LORETA} = \left(W^T W\right)^{-1} G^T \left(G\left(W^T W\right)^{-1} G^T + \lambda I\right)^{-1} \\ W = (\Omega \otimes \mathbf{I}_3) B^T B (\Omega \otimes \mathbf{I}_3) \end{gathered} \tag{5.33}$$

where B is a discrete spatial Laplacian operator. When given the task of localizing two simultaneous point sources (one deep and one superficial) within the same simulation study, LORETA achieved superior source localization accuracy when compared to MNE and wMNE. The minimum-norm approaches were unable to reproduce the deep source, while LORETA yielded a highly blurred reconstruction [65].

5.2.2.3 Standardized Solutions of Minimum-Norm Estimates

Another method applied to aid in the reconstruction of deep sources was the standardization of the MNE solution. The most popular accepted standardizations are the *dynamic statistical parametric mapping* (dSPM) method and *standardized low res-*

olution electromagnetic tomography (sLORETA) introduced by Dale et al. [19] and Pascual-Marqui et al. [66], respectively. While the solutions obtained from dSPM and sLORETA both take the form of the MNE solution $\hat{\boldsymbol{J}}$ standardized by the variance of the estimated current density $\boldsymbol{S}_{\hat{\boldsymbol{J}}}$, they differ significantly in their assumptions and the formulations of $\boldsymbol{S}_{\hat{\boldsymbol{J}}}$. dSPM assumes that the source of variation in the estimated current is solely from the measurement noise, thus:

$$\boldsymbol{S}_{\hat{\boldsymbol{J}}}^{dSPM} = \boldsymbol{M}_{MNE}(\lambda \boldsymbol{I}_N)\boldsymbol{M}_{MNE}^T \tag{5.34}$$

In contrast, the derivation of the variance for the estimated current density in sLORETA also takes into account variability from both the noisy measurement as well as variance within the actual source itself. This variance term is expressed as follows (see [66] for detailed derivation):

$$\begin{aligned} \boldsymbol{S}_{\hat{\boldsymbol{J}}}^{sLORETA} &= \boldsymbol{M}_{MNE}\left(\boldsymbol{G}\boldsymbol{G}^T + \lambda \boldsymbol{I}_N\right)\boldsymbol{M}_{MNE}^T \\ &= \boldsymbol{G}^T\left[\boldsymbol{G}\boldsymbol{G}^T + \lambda \boldsymbol{I}_N\right]^{-1}\boldsymbol{G} \end{aligned} \tag{5.35}$$

Finally, the standardized current density for dSPM and sLORETA is computed as:

$$\frac{\left(\hat{\boldsymbol{J}}_i\right)^2}{\left[\boldsymbol{S}_{\hat{\boldsymbol{J}}}\right]_{ii}} \tag{5.36}$$

where $\hat{\boldsymbol{J}}_i$ is the dipole moment at the ith source and $\left[\boldsymbol{S}_{\hat{\boldsymbol{J}}}\right]_{ii}$ is the ith diagonal element of matrix $\boldsymbol{S}_{\hat{\boldsymbol{J}}}$. In comparison with dSPM and the minimum-norm solution, sLORETA is claimed to be able to achieve zero localization error for a point source in a noise-less environment and the lowest localization error in noisy environments [66]. However, it has been shown [41] that while standardized solutions are effective at lowering localization error, they generally result in much larger spatial dispersion than non-standardized solutions (minimum-norm) (see Fig. 5.5). Moreover, while sLORETA yielded best accuracy in a single point source, LORETA appeared to perform better in cases where multiple distinct sources were active [10].

5.2.2.4 Inverse Solution in the Bayesian Framework

Finally, shifting from the minimization of error and cost function, a separate approach to source localization can be found through probabilistic Bayesian methods. In this context, solving for the inverse problem seeks to identify the probability distribution of the source activation given the observed scalp potentials. This is known as the posterior probability and can be expressed as:

$$p(\boldsymbol{J}|\boldsymbol{V}) = \frac{p(\boldsymbol{V}|\boldsymbol{J})p(\boldsymbol{J})}{p(\boldsymbol{V})} \propto p(\boldsymbol{V}|\boldsymbol{J})p(\boldsymbol{J}) \tag{5.37}$$

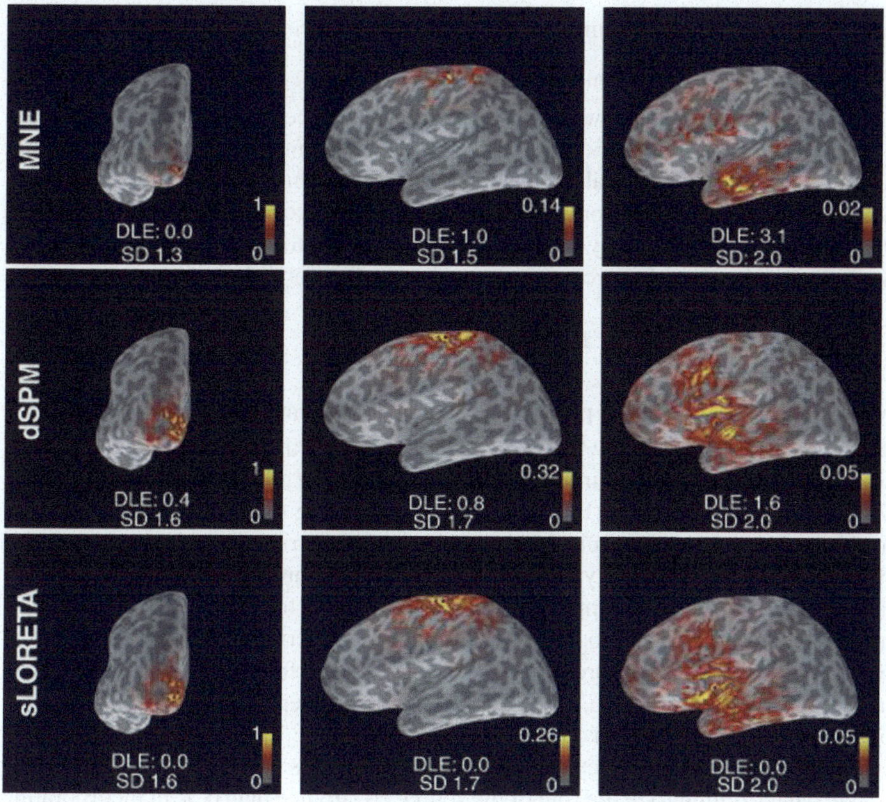

Fig. 5.5 Source localization performance of three EEG non-parametric inverse methods: MNE (top row), dSPM (middle row), and sLORETA (bottom row) at three source locations. The true location of the point-source is shown as a blue dot. Performance is evaluated in terms of dipole localization error (DLE) and spatial dispersion (SD). Figure reproduced from [41]

Again, assuming that source activity and noise are normally distributed with zero means and the respective covariance matrices $\boldsymbol{R}$ and $\boldsymbol{C}$, the likelihood and prior can be written as:

$$p(\boldsymbol{V}|\boldsymbol{J}) \propto \exp\left\{-\frac{1}{2}(\boldsymbol{GJ}-\boldsymbol{V})^T\boldsymbol{C}^{-1}(\boldsymbol{GJ}-\boldsymbol{V})\right\}$$
$$p(\boldsymbol{J}) \propto \exp\left\{-\frac{1}{2}\boldsymbol{J}^T\boldsymbol{R}^{-1}\boldsymbol{J}\right\} \tag{5.38}$$

where the solution can be obtained as the *maximum* a posteriori (MAP) estimate, in the form similar to that of (5.26) [7, 19, 35]: $MAP\left(\hat{\boldsymbol{J}}\right) = \|\left(\boldsymbol{G}\hat{\boldsymbol{J}}+\boldsymbol{n}\right)-\boldsymbol{V}\|^2+\lambda f\left(\hat{\boldsymbol{J}}\right)$, where $f\left(\hat{\boldsymbol{J}}\right)$ is the energy function associated with the prior, which can be interpreted identically to the $\boldsymbol{L}\left(\hat{\boldsymbol{J}}\right)$ term presented above.

Extending on this framework, Friston et al. introduced an EEG inverse method called Multiple Sparse Priors (MSP) [27, 28] which formulates the forward problem using a parametric empirical Bayesian approach on a hierarchical model [27–29]. MSP allows for the incorporation of multiple spatial covariance components into the data model, weighted by a set of hyperparameters to be estimated empirically. Specifically, the spatial source covariance matrix $\boldsymbol{R}$ is formulated as:

$$\boldsymbol{R}(\lambda^{R}) = \sum \exp(\lambda_j^{R}) Q_j^{R} \tag{5.39}$$

where each Q_j^R represent a potential source spatial covariance matrix weighted by the set of hyperparameters λ_j^R. The source activity and the hyperparameters for the source covariance can be estimated simultaneously using an iterative algorithm of the Expectation Maximization (EM). However, considering the linear model discussed here, the EM scheme is equivalent to the use of a restricted maximum likelihood (ReML) estimation of the hyperparameters, and the source activity is subsequently determined using a MAP approach (for more detailed formulation, see [29]). The main advantage of this method lies in its capability to determine the best source spatial prior model from a set of source covariance components. In fact, the weighting matrix in the classical MNE, wMNE, and LORETA can also be represented by the expression of $\boldsymbol{R}$ in (5.39), and MSP was found to outperform MNE, wMNE, LORETA in both spatial and temporal accuracy due to the flexibility of its spatial priors [27, 28].

We wish to emphasize the role of the source covariance matrix $\boldsymbol{R}$ in incorporating any possible prior knowledge about the spatial distribution of the sources into the inverse model. Specifically, the presented formulation of the inverse problem in this fashion has allows for many multimodal integration techniques that incorporate into $\boldsymbol{R}$ the spatial information provided by other imaging modalities (e.g. magnetoencephalography (MEG), functional MRI, functional NIRS).

5.3 Multimodal Integration

We have now covered how EEG from the scalp can be used to calculate and reconstruct cortical activity. EEG, of course, is not the only modality capable of viewing the brain; other methods are frequently utilized to gain a more direct view of the functional activity. *Functional magnetic resonance imaging* (fMRI), *functional near-infrared spectroscopy* (fNIRS), and *positron emission tomography* (PET) have all been used to obtain images of cortical activity without the complex calculations and volume conduction problems that come with the EEG methods. Conversely, these methods also do not have the high temporal resolution of EEG and are orders of magnitude slower than the activity they attempt to detect. We then have many approaches with different properties, which recent research has sought to combine into singular techniques that are both fast and accurate. In the following sections, we will provide a brief background for the complimentary modalities that are combined with EEG and then dig further into some of the algorithmic techniques used. Please note that much of this research is still ongoing and under discussion; many of the methods presented here are not as established as those in the source localization section, so the discussion will seek to broadly discuss the major developments within the field. Moreover, many of these methods exist as possible alternatives to one another with no single dominant approach or overriding theory. This comes at the cost of some algorithmic depth, though appropriate papers and groups will be cited for the readers' reference.

5.3.1 MEG and EEG Combinations

While the fundamentals underlying their signal detection differ greatly, MEG and EEG both represent the neural activity originated from a common source—the electric current resulting from the activity of a population of neurons. In contrast to the EEG's measurement of scalp electrical potentials through the volume conductor, MEG instead records the magnetic fields generated by the currents (both primary and volume) associated with sources in the brain (which can again be modelled as current dipoles). The amplitude of the magnetic field measureable outside of the head is on the order of a few hundred femtotesla (10^{-15}). This extremely small magnetic field is detected by MEG sensors known as *superconducting quantum interference devices* (SQUIDs). SQUID sensors are highly sensitive to small fluctuations in magnetic field strength, and therefore require a specially shielded recording environment to attenuate any external magnetic fields. MEG measurement also generally comes at a much higher cost than its EEG counterpart - a cost that arises due to both the shielding of the equipment and the permanent helium-cooling required to maintain the superconductive property of the SQUID sensors. In practice, a MEG recording is almost always accompanied by simultaneous EEG recording, providing possibilities for both parallel and integrated data analyses.

Like EEG, MEG signals also represent the superimposition of all primary and volume currents induced by the current sources in the brain. Naturally, source imaging techniques are also employable and desirable for MEG signals. However, the generation of the forward model for MEG signals is distinctly different from that of EEG due to the fundamental differences signaling characteristics. First, an important feature of MEG measurement is its insensitivity to sources that are radially-oriented to the scalp—it is only capable of measuring signals from "tangential" sources. Briefly, this is due to the fact that magnetic coils placed parallel to the scalp will pick up magnetic fields that are perpendicular to the coil (since the magnetic flux through the coil is measured). Radial current sources produce a magnetic field that is parallel to the sensor because of their orientation, making them invisible to the coils (see [1] for further discussion of the effect of source orientation). Secondly, the magnetic field is not distorted by the conductive properties of tissues, since the magnetic permeability of biological tissue is similar to that of empty space. As a result, MEG signals are not affected by the volume conduction problem and the modelling of the volume conductor is not needed. A detailed description and formulation of the forward problem for MEG is provided in [37]. Methods pertaining to solving the inverse problem are similar for both EEG and MEG.

While insensitive to radial sources, the resistance of MEG measurements to the blurring and distortion effects of volume conduction allows for better spatial resolution and accuracy. On the other hand, EEG signals are capable of capturing both "radial" and "tangential" sources, yet the resolution suffers due to volume conduction problem. These features are clearly complementary, making methods for the data integration analysis of these two modalities highly desirable. Owing to the fact the MEG and EEG signals are generated by the same underlying current sources, the integration analysis of MEG and EEG is made straightforwardly by combining the forward models of each individual method. Let us define a vector $\boldsymbol{j}$ of the unknown dipole strengths of the current source, with the electric and magnetic lead field matrices of $\boldsymbol{G}$ and $\boldsymbol{B}$, respectively, and the vectors of the corresponding recorded scalp potentials $\boldsymbol{v}$ and magnetic values $\boldsymbol{m}$. The combined forward expressions is then written in a concatenated form as:

$$\begin{bmatrix} \boldsymbol{G} \\ \boldsymbol{B} \end{bmatrix} [j] = \begin{bmatrix} \boldsymbol{v} \\ \boldsymbol{m} \end{bmatrix} \tag{5.40}$$

A scaling procedure is implemented by normalizing the rows of $\boldsymbol{G}$ and $\boldsymbol{B}$ to their respective norms and performing the same scaling operation on the measurement vectors $\boldsymbol{v}$ and $\boldsymbol{m}$ [6]. The resulting linear system is presented as:

$$\boldsymbol{y} = \boldsymbol{A}\boldsymbol{j} + \boldsymbol{n} \tag{5.41}$$

where $\boldsymbol{y}$ is the measurement vector representing the combined normalized electric and magnetic values, $\boldsymbol{A}$ is the combined normalized lead field matrix, and $\boldsymbol{n}$ is the measurement noise vector. This formulation is analogous to the EEG forward problem presented above, such that the same inversion scheme, with the same limitations, discussed for EEG can be applied to this system. Using a computer simulation study, the EEG+MEG integration analysis was demonstrated to be able to achieve significantly superior localization performance (in terms of residual error and temporal accuracy) in comparison to the separate unimodal analyses of EEG and MEG [6]. Further, EEG + MEG analysis results showed the best correspondence with the spatial pattern of neural activity obtained by functional MRI results [77] (see Fig. 5.6).

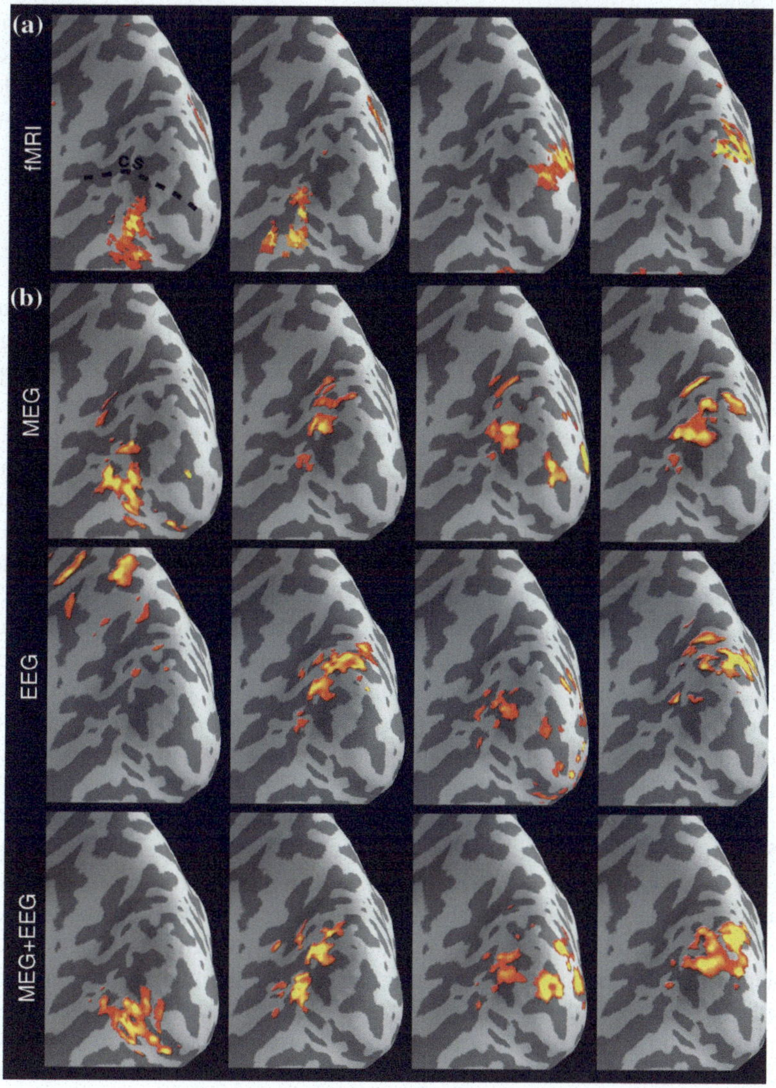

Fig. 5.6 Source localization results from unimodal EEG, MEG, and multimodal integration of EEG and MEG, in comparison with activation map obtained from fMRI data. MEG + EEG source analysis provided the best correspondence with fMRI results. Figure reproduced from [77]

5.3.2 *FMRI and EEG-FMRI Combinations*

While the combination of EEG and MEG can provide for effective source localization, researchers often prefer to combine EEG with modalities that feature a greater inherent spatial resolution, such as fMRI. To understand the nature of EEG-fMRI imaging, it will first be necessary to understand the nature of 4D magnetic resonance imaging; fMRI is a complex imaging modality that requires some investigation to appreciate, especially when contrasted with the more direct and simple EEG detection. For this, we will not be focused on the actual electrical activity of the brain, but will instead address the blood circulating within the cortical tissues.

5.3.2.1 The Basics of FMRI Acquisition and Analysis

When neurons fire within the brain, large amounts of energy are necessarily consumed by cells in the process of regulating and restoring their ionic gradients. The energy for this is provided in the form of blood borne glucose and oxygen, whose consumption creates a local decrease in oxy-hemoglobin. This decrease, paired with the local release of neurotransmitters, is perceived by astrocytes in the area of activity. The astrocytes then alter the blood flow of the brain, creating a sharp influx of oxygenated to compensate for the deficit caused by functional activity. This association between neural activity and cerebral blood flow is known as *neurovascular coupling* (NVC). The compensatory increase in local oxygen then serves as the basis for fMRI imaging.

fMRI takes advantage of this coupling, along with the fact that oxygenated and deoxygenated blood present with very different magnetic properties, to provide an indirect depiction of cortical activity. A series of magnetic fields are applied and pulsed across the sample or subject to rotate and displace the dipole moments of blood-borne hemoglobin. As these displaced dipoles return to their original states, the extra energy imparted by the magnetic field is released as a radiofrequency signal that travels unhindered through the skull, scalp, and open air. The detection of this signal is interpreted to produce greyscale images of the whole brain, one slice at a time. Statistical analyses are typically applied to identify which specific regions (in the form of voxels) show statistically significant differences between conditions. The result of this process is an image that is slow and indirect, but very spatially accurate. This forms a complimentary imaging modality to EEG, which is both fast and direct in its depiction of cortical activity but struggles with spatial blurring and inaccuracy.

While it is not considered a primary focus of interest, it will be worthwhile to spend some extra time discussing the primary statistical methods of fMRI. When MRI data is acquired, it comes in the form of greyscale 3D volumes. Anatomical scans will consist of a single high resolution volume while functional scans will consist of multiple lower resolution images taken at each timepoint during the scan. Each image will be made up of a series of 3D voxels, which are analogous to the pixels measured on 2D displays. Over the course of the scan, the intensity of each voxel will fluctuate both due to noise and actual shifts in the magnetic properties of the underlying tissues. A *general linear model* (GLM) is then constructed to determine the statistically significant changes in greyscale value that can be attributed to changes in the experimental condition. Following the simple, ideal form of the GLM, the intensity of a voxel (Y) should be linearly related to the condition (X) following the simple formula:

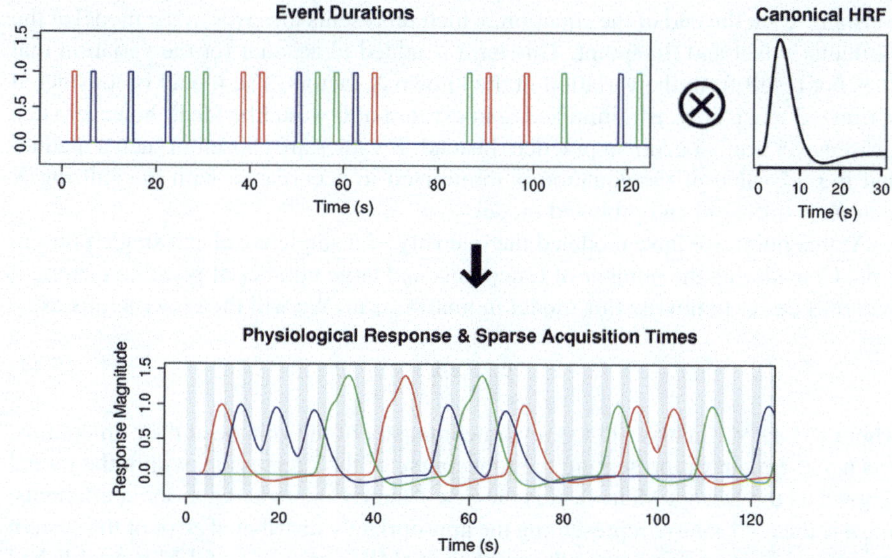

Fig. 5.7 Formulation of the estimates to the BOLD response. Here, the regressors for different theoretical events (indicated by event color) is convolved with the hemodynamic response function (HRF). Note the summation that occurs when events repeat quickly. Figure modified from [69]

$$Y = XB \tag{5.42}$$

where X is a representation of your experimental condition (for example, the X may be set up so that 1 represents an active task while 0 represents an inactive task) and B is a static coefficient. While this is an elegant representation, it is overly simple; cortical oxygenation does not operate in such a straightforward manner. The cerebral blood flow instead follows a specific pattern, known as a hemodynamic response function (HRF). To better match this cortical response, a numerical representation of the condition is convolved with the HRF, resulting in an elongated waveform. An example of this convolution can be seen in Fig. 5.7, where the HRF is convolved with small blocks of stimuli, whose color represents event type or task.

This gets us closer to a realistic model, but it is still insufficient—considering the complex setting of MRI recordings, a variety of factors outside of the experimental conditions may be seen to influence the voxel intensity. These can include a variety of directional or rotational movements, changes in the baseline voxel value, and even random error. Thankfully, we can expand the linear model to include each of these factors:

$$Y = X_1B_1 + X_2B_2 + X_3B_3 \ldots X_jB_j + e \tag{5.43}$$

where each X represents each condition or factor (convolved with the HRF) and B represents the corresponding coefficient for each factor. Note the addition of a constant "e" at the end of the equation, which represents the error of the model at the particular voxel and timepoint. This term is added to account for the variation that does not result from the variation in the known X factors. The model is considered optimized when error is minimized—an e-value of 0 would be ideal; however such a case never really occurs in practice. Instead, B values are chosen in such a manner that the 'e'-value of the equation is minimized in accordance with the shifting X factors, resulting in an optimized model.

At this point, we have modeled the intensity of a single voxel at a single point in time. Considering the number of timepoints and large number of possible factors, it becomes easier to rewrite this model in matrix form. We will then rewrite this as:

$$\boldsymbol{y} = \boldsymbol{x\beta} + \boldsymbol{e} \tag{5.44}$$

where y is a $t \times 1$ matrix of the greyscale intensity of the volume at the t timepoints, x is a $j \times t$ matrix representing the timecourse of the j regressors within the model (known as the Design Matrix), β is the $j \times 1$ matrix of static regressor coefficients, and e is the $t \times 1$ matrix representing the appropriately distributed error of the system at any given time point. This is the essence of fMRI analysis—a GLM is established and T- or F-tests are applied to identify specific voxels or regions of activity that are significantly different from their baseline value. An example of this construction can be seen in Fig. 5.8.

For many years, MRI stood as a singular imaging platform that was unable to be a part of simultaneous data acquisition due to the use of strong magnetic fields. More recent technological advances, however, have been enabling simultaneous EEG-fMRI. This advent of MRI-capable EEG devices has brought with it the search for algorithmic methods to capitalize on the temporal features of EEG and spatial features of fMRI. At the time of writing, a number of methods are available which can be broadly split into *Asymmetrical Methods* and *Symmetrical Methods* depending on how they treat the information from each modality. A broad overview of the topic can be seen in Fig. 5.9.

5.3.2.2 Asymmetrical Methods

Asymmetrical Methods were among the first devised which, in accordance with their name, place a stronger value on one modality over another. This typically takes the form of one modality generating the actual observed results while the secondary method provides a guiding influence. Once more, these approaches will split into categories to allow for a clearer analysis investigation.

fMRI-Constrained EEG

fMRI-informed EEG is perhaps the most fundamental of the multimodal approaches. Under this architecture, static fMRI maps are used to constrain EEG source localization results. The most straightforward method of combination is to restrict the source

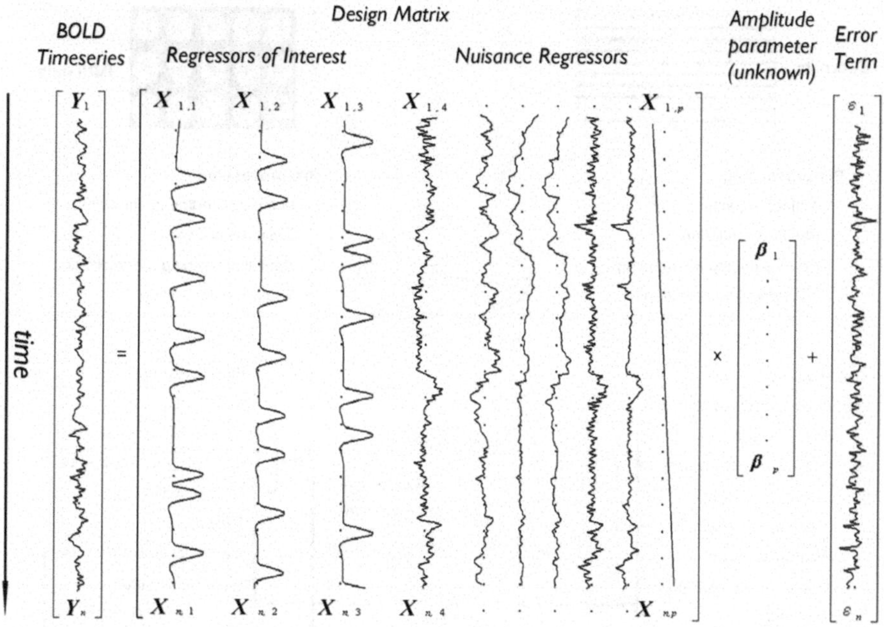

Fig. 5.8 Graphical representation of the general linear model method of modelling fMRI BOLD responses to the possible regressors. Figure reproduced from [58]

space so that dipoles are only placed at the MRI-active regions. This limited dipole seeding is less popular, however, as it no longer represents a distributed source model and again faces the problems seen by dipole fitting methods. An alternative approach, proposed by Anders Dale and Martin Sereno [17], applied a direct constraint from MRI on the distributed minimum-norm solution. Recall that the equation for the classical minimum-norm estimates (see (5.29) above):

$$\boldsymbol{M} = \boldsymbol{R}\boldsymbol{G}^T\left(\boldsymbol{G}\boldsymbol{R}\boldsymbol{G}^T + \lambda\boldsymbol{C}\right)^{-1}$$

where $\boldsymbol{M}$ is the optimal linear inverse operator and $\boldsymbol{R}$ and $\boldsymbol{C}$ are again the respective source and noise covariance matrices, which are set to identity matrices in classical MNE. The multimodal constraint in this case will be applied directly on the source covariance matrix $\boldsymbol{R}$, changing the weight of each source according to whether or not it is within an fMRI-active region. This method requires experimenters to establish both the threshold at which sources will be considered fMRI-active (typically in α, p, or q-values) as well as the weight to apply each to each individual source. Common application of this method sets the p or q threshold for the fMRI map at 0.05 with the diagonal term of $\boldsymbol{R}$ for the significantly activated sources receiving a source covariance value of 1 and non-significant sources receiving a source covariance value of 0.1 [49], the off-diagonal terms are set to zeros, reflecting a lack of hemodynamic coupling to other cortical sources. On the other hand, Babiloni et al. [6] introduced a

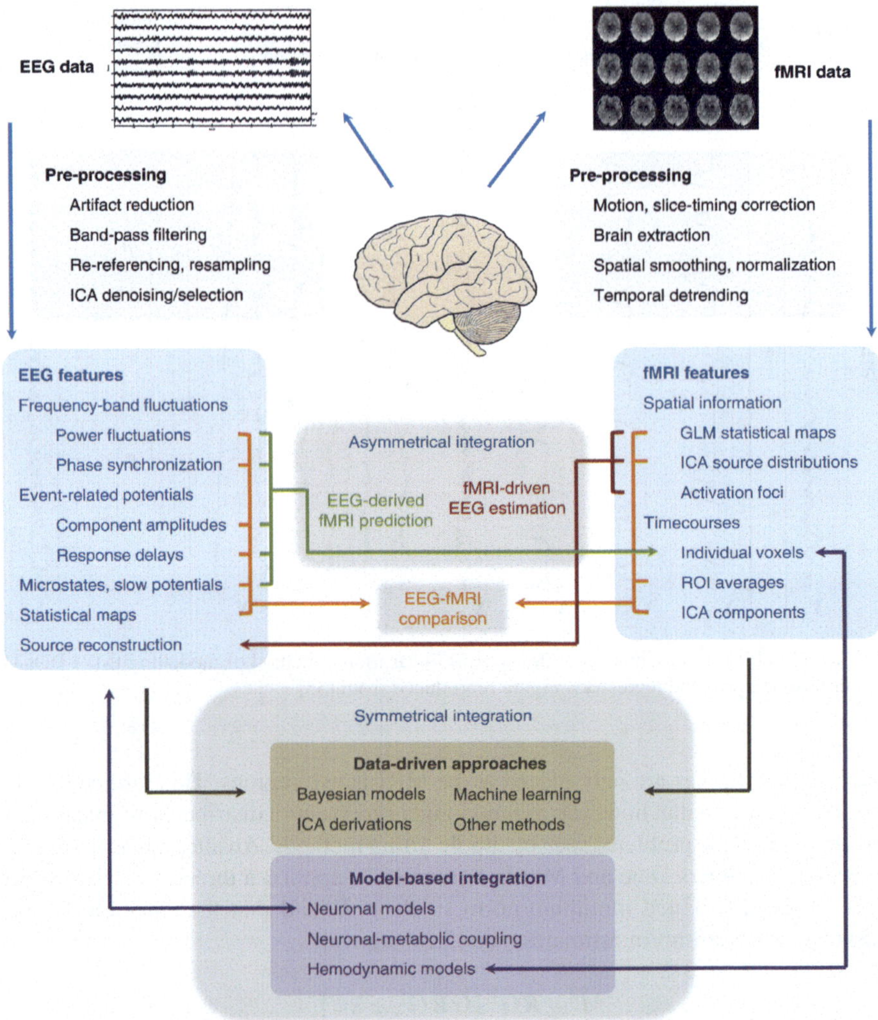

Fig. 5.9 An overview of different data analysis schemes for the multimodal integration between EEG and fMRI data. Figure reproduced from [44]

formulation of $\boldsymbol{R}$ that describes the spatial coupling of the hemodynamic responses of cortical sources by modelling the off-diagonal terms in $\boldsymbol{R}$ as proportional to the correlation found in the fMRI time-courses of the corresponding pair of cortical sources. In any case, fMRI-informed EEG source localization results in significant improvement in localization accuracy over EEG-only inversion methods [6, 49].

The result of the above method is very similar to the typical MNE-based source localization, applied on the experimenter's model of choice. Completed reconstruction has a similar temporal scale and rapid detection, with some accompanying uncertainty in the calculation of the cortical current sources. The fMRI activation map, in this case, serves only to restrict the potential activity of sources in the cortical layer by penalizing sources in fMRI-inactive regions. This is emblematic of asymmetrical methods; one imaging modality provides the directly observed results while the remaining method acts as a guiding influence. There are also a few considerations which must be made for this fMRI-MNE method. First, strong activity in the cortical reconstruction can overcome the penalty imposed by the fMRI map. This means that the erroneous activity of sufficient strength may still be present in the results. Second, the MRI constraint does not interact with other noise-normalized estimates (dSPM, sLORETA, etc.). While MNE itself aims to provide a reconstruction of cortical currents, these noise-normalization algorithms seek to identify where cortical activity is different from baseline noise. This function is essentially the same as fMRI, which identifies the 3D voxels that are significantly different from background noise. Thirdly, in the above implementation, fMRI spatial information acts as a "hard" constraints; it is assumed to be the absolute truth, in spite of cases where fMRI "extra" sources (sources deemed active in fMRI but not EEG, [49, 51]) can be found. Finally, there is a built-in assumption that the constraints provided by fMRI are applicable at each time instance of the EEG measurement. The fMRI activation map employed in this method is static, regardless of the time point use in analysis—this means that any erroneous activity at an MRI-active voxel will be amplified similarly to true activation. To alleviate the issue of temporal mismatch between fMRI and EEG, Liu and He [52] proposed an EEG inversion approach that utilized the fMRI information as time-variant spatial constraints. The fMRI-derived prior spatial weights are adaptively varied at each time instance of the EEG time-course based on all the EEG single-trials before averaging. On the other hand, Daunizeau et al. addressed the fMRI "hard" constraint issue by dividing the fMRI activation map into multiple submaps and estimating the optimal subset of fMRI weights using model comparison approach in the parametric empirical Bayesian framework [20]. fMRI spatial information can be utilized in a spatiotemporal specific fashion [60], employing an appropriate subset of the fMRI activation map as spatial priors for a sliding-window of EEG time segments, thus improving the spatial and temporal accuracy in complex and dynamic brain activity [61].

EEG-Informed fMRI

The above method has shown a common combination, wherein fMRI is used as a basis to constrain EEG-based source localization. The lofty temporal resolution of this approach makes it ideal for investigative or computational approaches. Oftentimes, however, clinical and scientific researchers may instead wish to highlight or localize a specific feature or spike in EEG activity. In these cases, they may wish to rely more heavily upon the high spatial resolution of fMRI than the temporal speed of EEG. This can be particularly valuable when the events targeted for characterization fall into one of the following categories: (1) events that are largely uncontrolled or unpredictable (investigations of epileptic activity); (2) events with a large degree of variability; and (3) events that are invisible to fMRI alone. Serving as an converse method to fMRI-informed EEG, results obtained by EEG-informed fMRI are similar to results from MRI alone; they are largely static and comprised of a small number maps (or a single map) of activated 3D voxels. As with fMRI, results are not subject to the volume conduction problem or intense calculation and are not susceptible to the error that may arise from EEG source localization. While the lack of temporal resolution may lead researchers to overlook EEG-informed fMRI when attempting to characterize cortical activity, the robust nature of the MRI results and guiding effects of EEG have made this a clinically valuable and worthwhile imaging approach.

When attempting to perform EEG-informed fMRI, we must recall that our primary data method of choice is fMRI. Whereas fMRI-informed EEG built results from the fMRI into the EEG source localization framework via the source covariance matrix, we will turn once more to the General Linear model presented in the section on fMRI analysis. Recall the model:

$$\boldsymbol{y} = \boldsymbol{x}\boldsymbol{\beta} + \boldsymbol{e}$$

where $\boldsymbol{y}$ is a $t \times 1$ time-course of bold activity for every voxel within the MRI volume, $\boldsymbol{x}$ is an $r \times t$ matrix representing the timecourse, t, of the r regressors within the model, $\boldsymbol{\beta}$ is the $r \times 1$ matrix of static regressor coefficients, and $\boldsymbol{e}$ is the $t \times 1$ matrix representing the error of the system at any given time point. Recall further that the r regressors used within the model are a series of $t \times 1$ signals used to model the status of the various factors that can directly impact voxel intensity—motion, drift, noise, stimuli of interest, etc.—which are ultimately convolved with general or subject-specific models of the cortical hemodynamic response. Logically, an EEG signal of the activity of interest can be incorporated directly into this model as one of these regressors. Integration of this type may require some extra steps during this processing, as EEG signals would need to be downsampled to match the timescale of collected fMRI. Once appropriate processing has been performed, however, this incorporation becomes straightforward and requires only that the EEG data of interest be added to the x matrix as a regressor of interest for the fMRI statistical analysis. More pressing questions, then, are exactly what EEG feature is selected and how that selection is performed-questions that will fall to individual researchers.

5.3.2.3 Symmetrical Methods

At this point, we have discussed the predominate asymmetrical methods for combining EEG and fMRI. These methods have been used broadly and have significantly impacted the field of biomedical imaging. Reviewing these applications and their biases, however, it becomes clear that the multimodal integration is only implemented as an extension of existing unimodal methods. It is natural that continued development would focus instead on novel methods that better utilize both methods. The inherent differences between EEG and fMRI have made this advancement difficult, however, requiring more mathematically advanced methods to achieve these tasks. Unlike the EEG-informed fMRI and fMRI-constrained EEG, which each have a standard approach, symmetrical methods have yet to present singular methods for discussion. It may be better to explore this topic in a historical manner, then, reviewing major landmark developments. Once more, however, we can simplify the topic by splitting it into two major categories: data driven approaches and model-based methods.

Data-Driven Integration

Data-driven methods are those that focus predominately on integrating the datasets from EEG and fMRI. Under these paradigms, little attention is given to the actual structures and dynamics of neural populations or how they interact with cortical blood flow. It is instead assumed that the numerical methods will provide the necessary context for application. This means that the methods are generally straightforward and computationally accessible.

Perhaps the earliest effort to appear as a symmetrical data-driven approach was presented by Martinez-Montes et al. in 2004 [55]. Under their approach, a multiway Partial Least-Squares (PLS) method was used to decompose EEG and fMRI data into a sum of elements, which were called "atoms." Following this method, EEG atoms contained three main properties—their spatial, temporal, and spectral signatures—following a method of Parallel Factor Analysis, where the EEG Signal was represented as a trilinear model of these parts:

$$S_{dwt} = \sum_{k=1}^{N_k} a_{dk} b_{wk} c_{tk} + \varepsilon_{dwt} \tag{5.45}$$

where S is the EEG signal; a, b, and c are normalized vectors influenced by d, w, and t, which respectively represent electrode (spatial), frequency (spectral), and time (temporal) components; N_k is the number of components; k specifies the current component; and ε is the error. fMRI is similarly modeled considering time and voxel as their key dimensions:

$$F_{st} = \sum_{k=1}^{N_k} u_{sk} v_{tk} + \varepsilon_{st} \tag{5.46}$$

where F_{st} is the fMRI signal; u and v represent the two signatures defined by their voxel (spatial component represented by s) and time (temporal component represented by t). Once more, N_k and k are the number of components and the current component, and ε is the error. These representations can be seen in Fig. 5.10.

The method then sought to maximize the covariance between the temporal properties of the EEG and fMRI signals (the $\boldsymbol{c}$ and $\boldsymbol{v}$ vectors in Fig. 5.10), matching the signals. EEG signals were further broken down into frequency bands of interest, focusing on only one band at a time and identifying which frequencies presented a significant correlation with the fMRI signal and how they were distributed throughout the brain. As in EEG-informed fMRI, this integration required that EEG data be downsampled and convolved with an appropriate HRF to exist within the same

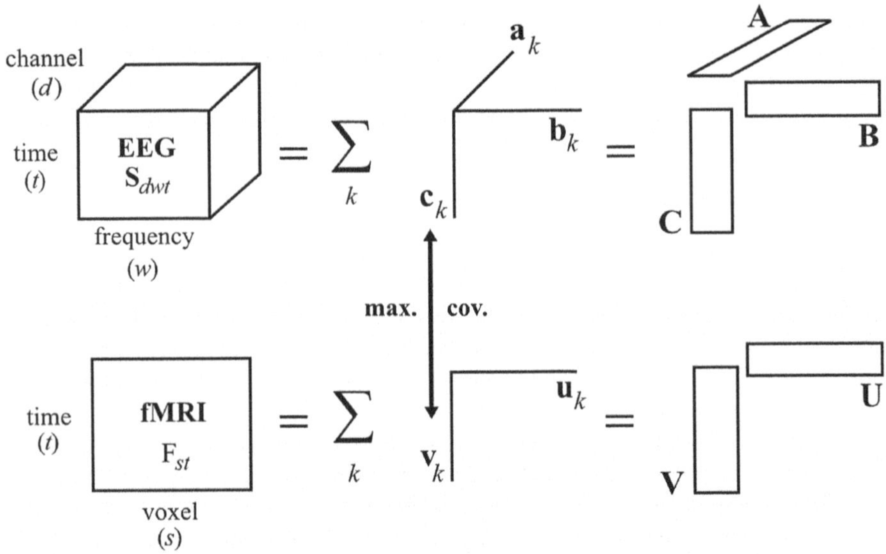

Fig. 5.10 A depiction of how EEG and fMRI data sets are represented as the sum of their constituent signatures and subsequently broken down in the PLS method. EEG data features three signatures—time, frequency, and channel—while fMRI data has two—time and voxel. The data sets are then temporally aligned by maximizing their temporal covariance. Subsequent analysis here involved selecting individual frequencies and identifying the significant correlations between the respective spatial dimensions of the EEG and fMRI signals. The figure is reproduced from [55]

temporal scale as the fMRI data. A LORETA-based source localization algorithm was used to provide the spatial signature for each EEG atom for covariance analysis (note: LORETA itself could not be used as EEG atoms were defined by spectral power and not voltage). While the algorithm did face limitations when accounting for spatial blurring, uniformity, and the interactions between the three properties of interest, it served as a noteworthy advancement and one of the first examples of *multimodal data fusion*. Methods like these would become more common as the field moved forward.

A similar, more ERP-oriented method known as Joint ICA was presented shortly after as another fusion method [15, 59]. To do this, ERPs were first derived from scalp EEG signals following an auditory oddball experiment. A single focal scalp electrode (Cz in their case) was chosen for the fusion analysis along with the fMRI data. The two data sets were represented as the respective generative models following the infomax principle [9]:

$$\boldsymbol{x}^E = \boldsymbol{A}\boldsymbol{s}^E \quad \text{and} \quad \boldsymbol{x}^F = \boldsymbol{A}\boldsymbol{s}^F \tag{5.47}$$

where $\boldsymbol{x}$ and $\boldsymbol{s}$ respectively represent the mixed data from each subject and the source as obtained by either fMRI ($\boldsymbol{x}^F$, $\boldsymbol{s}^F$) or EEG ($\boldsymbol{x}^E$, $\boldsymbol{s}^E$), and $\boldsymbol{A}$ is a shared linear mixing matrix. Data vectors can then be formed for each subject as:

$$\boldsymbol{x}_i = \left[\boldsymbol{x}_i^F \boldsymbol{x}_i^E\right] \text{ and } s_i = [s_i^F s_i^E] \tag{5.48}$$

where i indicates the data from the i-th subject. A resulting update equation is used to compute a shared unmixing matrix and the fused ERP fMRI sources ($\boldsymbol{u}^F$ and $\boldsymbol{u}^E$) as:

$$\Delta \boldsymbol{W} = \eta\left\{\boldsymbol{I} - 2\boldsymbol{y}^E\left(\boldsymbol{u}^E\right)^T - 2\boldsymbol{1}^F\left(\boldsymbol{u}^F\right)^T\right\}\boldsymbol{W} \tag{5.49}$$

where $\boldsymbol{y}^E = g\left(\boldsymbol{u}^E\right)$, $\boldsymbol{y}^F = g\left(\boldsymbol{u}^F\right)$, and $g(x) = 1/(1 + e^{-x})$, which represents the nonlinearity of the neural network. Independent components could then be isolated from the fused data sets and used to identify important motifs and patterns of brain response. It would then be up to the experimenter to determine which components are important or meaningful based on their own criteria and evaluation. Localization within the source paper was then performed for the N spatial and temporal components by rewriting them as:

$$\boldsymbol{T} = [\boldsymbol{t}_1 \ldots \boldsymbol{t}_N] \text{ and } \boldsymbol{S} = [s_1 \ldots s_N] \tag{5.50}$$

where $\boldsymbol{t}_i$ is a $T \times 1$ vector of the T timepoints and $\boldsymbol{s}$ is a $V \times 1$ vector of the V brain voxels within the MRI space. An overall fMRI movie ($\boldsymbol{M}_F$) and ERP timecourse ($\boldsymbol{M}_E$) were then calculated using the respective equations of:

$$\boldsymbol{M}_F = |\boldsymbol{T}| \times \boldsymbol{S}^T \text{ and } \boldsymbol{M}_E = \boldsymbol{T} \times |\boldsymbol{S}|^T \tag{5.51}$$

The result of this procedure is a technique that combines both EEG and fMRI into a single, joint data space that accounts for the independent features of both modalities. Unlike the PLS method, neither modality here is treated as dependent or independent and, thanks to the use of MRI for localization, the method does not rely on cortical models with fixed numbers of dipoles or potentially blurred calculation through cortical layers. On the other hand, the method only accounts for EEG data from a single focal electrode and does not incorporate the larger spectrum of data collected throughout the scalp. Joint ICA also requires that data are fitted to the maps obtained through fMRI, and any misalignment between theses may cause errors in the results or interpretation. Finally, while this is useful for specific ERP analysis, it is unable to address the broader context of EEG without serial calculation. This may pose issues for those seeking to understand brain connectivity or the dynamic cortical activity that may be incurred during complex tasks. In practice, Joint ICA has been used to explore brain activity in schizophrenia [26] and during error-monitoring tasks [22].

Moving away from joint feature spaces, Bayesian methods represent another interesting direction for symmetrical integration. The first landmark method for this integration was proposed in 2007 and made use of a generative model in somewhat of a similar manner to that seen in the Joint ICA [21]. The model in this case jointly incorporated EEG/fMRI sources as unknown hierarchical priors within a Variatonal

Bayesian learning scheme, resulting in an estimate of the common spatial profile for the modalities. Included as a part of this approach is an estimation of the spatial structure of the EEG/fMRI coupling and uncoupling. As with EEG-only data, probabilistic approaches are of great interest because they are robust to imprecision and accommodate a degree of error without greatly inhibiting localization ability. Returning briefly to the simplified form of Bayes' rule discussed earlier, we have:

$$p(\boldsymbol{J}|\boldsymbol{V}) = \frac{p(\boldsymbol{V}|\boldsymbol{J})p(\boldsymbol{J})}{p(\boldsymbol{V})}$$

where $p(\boldsymbol{J}|\boldsymbol{V})$ is the probability density function (pdf) of $\boldsymbol{J}$ given the data $\boldsymbol{V}$, $p(\boldsymbol{V}|\boldsymbol{J})$ is the data likelihood, $p(\boldsymbol{J})$ is the prior pdf of $\boldsymbol{J}$, and $p(\boldsymbol{V})$ is the data "evidence". Unwritten in this formulation are a series of other hyperparameters that control the distributions of these probabilities. An expanded form of Bayes' rule can be written as:

$$p\left(\boldsymbol{J}|\sigma^2, \epsilon^2\boldsymbol{V}, H\right) = \frac{p\left(\boldsymbol{V}|\boldsymbol{J}, \sigma^2\right)p(\boldsymbol{J}|\sigma^2, \epsilon^2, H)}{p(\boldsymbol{V}|\sigma^2, \epsilon^2, H)} \tag{5.52}$$

where σ^2 and ϵ^2 represent a set of mutually independent hyperparameters and H is an undefined hypothesis. In the absence of fMRI data, the hypothesis H is established as an uninformative prior, represented by using an identity matrix $\boldsymbol{I}_n$ as a prior covariance matrix:

$$\boldsymbol{J}_j \sim N\left(0_n, \frac{\sigma^2}{\epsilon^2}\boldsymbol{I}_n\right), \quad j = 1, \ldots t \tag{5.53}$$

Knowing the fMRI activation map, **Z**, we can instead introduce it as an informative hypothesis wherein the source intensities at time j are a function of that fMRI map.

$$\boldsymbol{J}_j \sim N\left(0_n, \frac{\sigma^2}{\epsilon^2}f(\mathbf{Z})\right), \quad j = 1, \ldots t \tag{5.54}$$

Following this model, the source localization results are not directly dependent on the EEG alone or the fMRI, instead relying on factors from each to control the probabilistic model. This creates a favorable interaction between the unknowns of each method without relying too heavily on either or making excessive assumptions regarding the correctness of either.

The development of these methods served as the advent of data-driven symmetrical EEG-fMRI integration. These approaches do not account for all of the development within the field; however, as existing approaches are continually refined and new methods are brought forth. The techniques that we have presented here serve only as major landmarks within the field, which can help us both understand and develop newer integration algorithms. For example, one later-developed method used ICA to isolate the temporal and spatial components from EEG and fMRI separately, which was then linked using an Empirical Bayesian model [46]. This, in essence, performs both of the previous asymmetric methods and uses a Bayesian approach to fuse them as a newer combination. Alternative methods have extracted EEG and fMRI features (BOLD and ERP peak latency and amplitude, BOLD percent signal change, RMS measure, etc.), establishing the probability distribution of each and using these to determine the information shared by the components [63]. Others have created spatially adaptive priors for use within Bayesian frameworks, developed by implementing measures of Total Variation [53]. Finally, some have sought to introduce measures of graph theory or connectivity into the framework. An early method for this performed functional network analysis on the fMRI and EEG signals using spatial ICA and Granger Causality, again linking these within an Empirical Bayesian Framework [47]. The important message to derive from all of these is not just the individual approaches, but also the numerous ways that they can be combined, adjusted, or refined to improve on current technology. Though much has been accomplished, it is certain that many more methods will be developed with the potential to both advance multimodal imaging and translate across fields.

Model-Based Approaches

Model-based approaches, on the other hand, are founded upon the development of realistic models of the biological and physical factors that give rise to the detected BOLD and EEG signals [71]. In general, this means that data-driven approaches are simpler and better suited to naïve cases while model-driven approaches are more conceptually complete and informative, albeit with an increased computational cost. It also means that model-based approaches require a somewhat deeper understanding of the factors connecting EEG to fMRI. For example, we have established that neurovascular coupling as the underlying mechanism that connects the EEG and fMRI modalities. This coupling is not constant, however—decoupling between the BOLD and EEG signals can be informative depending on the situation. Decoupling has been observed in a variety of specific cases, including decoupling in the frontal lobe during locomotion and in cases of cerebral amyloid angiopathy [67] or seizure [75]. An ideal model will also need to deal with the dynamics of whole neuronal populations and once again address how these dynamics are represented at the scalp level. These populations may show complex activity, including both inhibitory and excitatory interactions within context of a population firing pattern that may be conditionally rhythmic or transient. True neurophysiologic models are then difficult to generate; the core principles for them are highly complex and any errors will be amplified as the model is built outward from the cell populations and their properties [71]. Thankfully, our current discussion has provided a firm basis in EEG source localization and reconstruction, which accounts for one major aspect of neurogenerative modelling.

The general pattern of the forward model and inverse problem, as explored above, reappears here. Forward models will serve to represent the processes that generate EEG and fMRI data while the inverse calculation will identify the model conditions responsible for observed data [42]. Early single and double columnar models of neuronal population activity were generated following a biological representation of excitatory pyramidal cells modulated by inhibitory interneurons and excitatory stellate cells [43]. This method proved to be effective at reproducing a variety of oscillating and spiking patterns that closely resembled experimental data. The model has since been expanded to cover a single or multiple cortical areas with the added context of hemodynamic coupling [4, 5]. While these models have shown a remarkable amount detail, it will be easier to illustrate the overall process of neurogenerative modelling using simpler approaches, such as that presented by Buxton et al. in 2004 [14], which modeled the total neural activity as the difference between excitatory and inhibitory inputs in the following equations:

$$N(t) = s(t) - I(t) \tag{5.55}$$

$$\frac{\mathrm{d}I}{\mathrm{d}t} = \frac{\kappa N(t) - I(t)}{\tau_1} \tag{5.56}$$

where $N(t)$ is the neural activity, $s(t)$ is the excitatory input, $I(t)$ is the inhibitory input, κ is a gain factor, and τ_1 is a time constant. Other models have also been developed and presented, including models specific to cortical regions [34], though there has yet to be a broadly accepted model for the collective activity of neuronal populations [11].

Having modeled the neural population of interest, the next aspect to be addressed is the connection from this population to the associated hemodynamics. As an example of this, we will turn to a model generated by Buxton et al. that describes the vascular BOLD response as a balloon fed by a vascular compartment [13, 14]. Following this model, two main variables—the total deoxyhemoglobin ($q(t)$) and volume of the compartment (v)—are described as:

$$\frac{\mathrm{d}q}{\mathrm{d}t} = \frac{1}{\tau_{MTT}}\left[f(t)\frac{E(t)}{E_0} - \frac{q(t)}{v(t)}f_{out}(v,t)\right] \tag{5.57}$$

$$\frac{\mathrm{d}v}{\mathrm{d}t} = \frac{1}{\tau_{MTT}}[f(t) - f_{out}(v,t)] \tag{5.58}$$

where τ_{MTT} is the mean transit time through the balloon at rest (~3 s), f is the bloodflow into the compartment, f_{out} is the bloodflow out, $E(t)$ is the O_2 extraction fraction at the time t, E_0 is the baseline O_2 extraction fraction, $q(t)$ is the basline-normalized deoxyhemoglobin, and $v(t)$ is the baseline-normalized *cerebral blood volume* (CBV). Considering the viscoelastic properties of venous walls, f_{out} can be further modeled as a function of the volume of the balloon compartment:

$$f_{out} = v^{\frac{1}{\alpha}} + \tau\frac{\mathrm{d}v}{\mathrm{d}t} \tag{5.59}$$

where τ is a time constant that can take different values during inflation and deflation and α is a constant used to describe the relationship between flow and volume at steady state (~0.4). Modeling the neurovascular coupling between these hemodynamic representations and the actions of neural populations rises as the next point of interest. Continuing with Buxton's model and assuming a neural activity of $N(t)$, the cerebral blood flow (CBF) and $CMRO_2$ are respectively modeled as

$$f(t) = 1 + (f_1 - 1)h\left(t - \delta t_f\right) * N(t) \tag{5.60}$$

and

$$m(t) = 1 + (m_1 - 1)g(t - \delta t_m) * N(t) \tag{5.61}$$

where f_1 and m_1 represent the normalized CBF and $CMRO_2$ responses, δt_f and δt_m are the relative delay of each response from the stimulus, * indicates convolution, and $g(t)$ and $h(t)$ are impulse response functions. The BOLD signal change can finally be modeled as:

$$\frac{\Delta S}{S_0} = A \cdot \left(1 - f^{\alpha-\beta} m^{\beta}\right) \tag{5.62}$$

where ΔS is the change in MR signal, S_0 is the resting signal, A is the maximum possible change in BOLD signal (contingent on MRI acquisition factors) and β is a constant related largely to MRI field strength (~1.5 for 1.5T or 3T machines).

At this point, we have followed a model beginning with a neuronal population and extending outward to influence both BOLD and EEG signals. While the approaches presented here have been important to the field, it is important to remember that neurogenerative modeling techniques are highly customizable and different models can be combined and altered in a number of ways to suit the case at hand. This flexibility and the explanatory power of the models make them valuable approaches within the field—particularly when attempting to verify or validate an existing hypothesis.

5.3.3 EEG-FNIRS

fNIRS can be seen as a relative of fMRI. While it does not utilize the strong magnetic fields of MRI, fNIRS provides researchers with a similar measurement of cortical hemodynamics. Signals in this case are created by near-infrared light from either laser diode or LED sources distributed throughout the scalp, which emit light between the wavelengths of 650 and 950 nm. Light within this range is able to transmit through the tissues of the head (cortex, skin, skull, scalp, CSF, etc.), though scattering from these intermediate tissues causes the near-infrared signal to curve through the head in a banana-shaped pattern. This allows the light achieve a probing depth of ~3 cm and return back towards the scalp surface, where separate optodes are placed to detect any changes in the signal intensity (see Fig. 5.11).

As with the magnetic field of MRI, the main focus for detection for fNIRS is hemoglobin, which serves as the main chromophore in blood and presents with different absorption spectra depending on whether or not it has bound oxygen. To

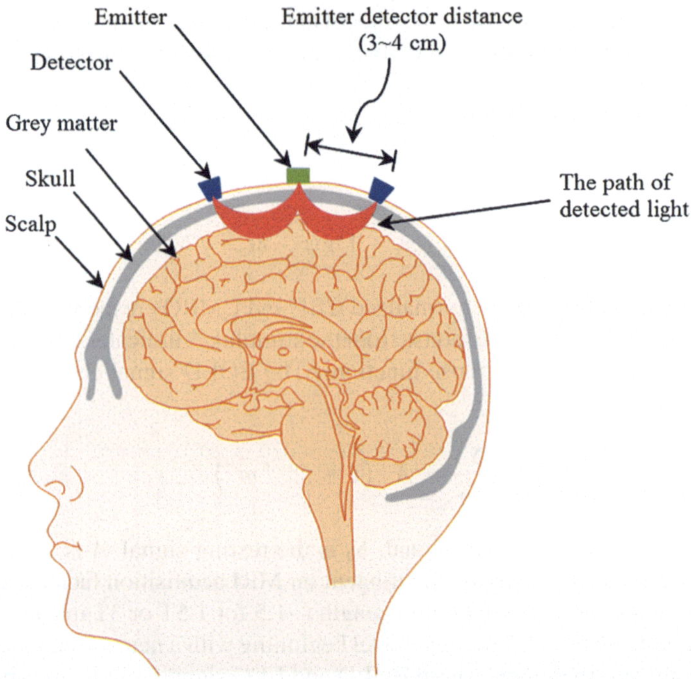

Fig. 5.11 A schematic of the signal acquisition in functional Near-Infrared Spectroscopy. Light from the source scatters through the brain tissue in a banana-shaped path before being picked up by the detector. Figure reproduced from [55]

suit this, NIRS systems will typically inject two or more wavelengths of light, with at least one in the range of 810–860 nm and one in the range of 710–770 nm. These will serve for the respective detection of oxy- and deoxy-hemoglobin in accordance with the absorption spectra for each chromophore.

Thanks to the use of scalp-based sources and detectors, fNIRS provides a portable and affordable method for detecting the same BOLD response observed by fMRI. The light-based signal is also resilient to motion noise and can make highly specific measurements of the BOLD signal, including non-relative measurements of the oxy, deoxy, and total hemoglobin within the cortex. On the other hand, while the signals of fNIRS can detect hemoglobin within the cortex, it does not provide any true anatomical detail for the signals. While signals are not as blurred as EEG and some locations can be inferred by fiducials, the resultant signal is most frequently seen as a scalp topography. Further, the light-based detection only reaches sources 2–3 cm deep and cannot measure deep-seated sources within the brain, making the detection of the insular or cingulate cortices a challenge. Finally, while fNIRS instrumentation may feature a high sampling rate, the underlying BOLD signal is still relatively slow and does not operate on the same temporal scale as cell populations. Regardless of the potential limitations of the hemodynamic detection, low cost, motion resilience, and portable nature of fNIRS have made it an intriguing counterpart to EEG, particularly in BCI applications [2].

As a relatively young modality, signal processing for fNIRS alone has been fairly simple. Methods have often included simple observation or T-tests/ANOVAs based on average signals. Fourier analysis has also been applied, assuming that hemodynamic peaks occurring with the same frequency as a stimulus should be attributable to the task. Statistical Parametric Mapping (SPM) and the GLM were also adopted for fNIRS in a similar manner to their use in fMRI. Full integration between EEG and fNIRS has yet to be fully explored, in part due to the relative youth of the technology. Current methods adopted procedures such as using the β coefficients do determine informative EEG channels for BCI classification [48], while others have derived the independent band powers from both EEG and fNIRS frequency bands for use in linear discriminant analysis and subsequent classification in a hybrid BCI [23]. Bayesian approaches have also appeared here, wherein fNIRS data is incorporated as a hierarchical prior [3]. Though the field is still young, we can see here the incorporation of methods established earlier (Bayesian methods, GLM-based analyses, etc.). These represent common themes that pervade many multimodal methods that will continue to serve EEG combinations in the future, even as new approaches and imaging modalities are developed.

5.4 Conclusion

EEG-based source localization and multimodal imaging stand as burgeoning topics within the field of biomedical imaging, and research into both topics has remarkable breadth and depth. New approaches are constantly arising in this field, pushing it further as more accurate (and often more complex) methods arise. We have sought here to provide a functional basis from which the general principles and seminal methods of source localization can be understood. This necessarily has driven through discussions of cortical modeling, types of models, and the forward and inverse calculations used to connect EEG to its potential sources. Continuing beyond this, we have explored a variety of the possible multimodal combinations that EEG features in and provided both algorithmic and practical examples for how these fusions are performed. It is important at this stage to remember that none of this discussion is comprehensive; even the topics explored more heavily in this chapter have not achieved their full depth. The readers are instead encouraged to look through the cited materials on their own, for many of the individual topics that we have touched on could be the subject of full books.

References

1. S.P. Ahlfors, J. Han, J.W. Belliveau, M.S. Hämäläinen, Sensitivity of MEG and EEG to source orientation. Brain Topogr. **23**(3), 227–232 (2010)
2. S. Ahn, S.C. Jun, Multi-modal integration of EEG-fNIRS for brain-computer interfaces–current limitations and future directions. Front. Hum. Neurosci. **11**, 503 (2017)
3. T. Aihara, Y. Takeda, K. Takeda et al., Cortical current source estimation from electroencephalography in combination with near-infrared spectroscopy as a hierarchical prior. Neuroimage **59**(4), 4006–4021 (2012)
4. A. Babajani-Feremi, H. Soltanian-Zadeh, Multi-area neural mass modeling of EEG and MEG signals. Neuroimage **52**(3), 793–811 (2010)
5. A. Babajani, H. Soltanian-Zadeh, Integrated MEG/EEG and fMRI model based on neural masses. IEEE Trans. Biomed. Eng. **53**(9), 1794–1801 (2006)
6. F. Babiloni, D. Mattia, C. Babiloni et al., Multimodal integration of EEG, MEG and fMRI data for the solution of the neuroimage puzzle. Magn. Reson. Imaging **22**(10), 1471–1476 (2004)
7. S. Baillet, J.C. Mosher, R.M. Leahy, Electromagnetic brain mapping. IEEE Signal Process. Mag. **18**(6), 14–30 (2001)
8. S.B. Baumann, D.R. Wozny, S.K. Kelly, F.M. Meno, The electrical conductivity of human cerebrospinal fluid at body temperature. IEEE Trans. Biomed. Eng. **44**(3), 220–223 (1997)
9. A.J. Bell, T.J. Sejnowski, An information-maximization approach to blind separation and blind deconvolution. Neural Comput. **7**(6), 1129–1159 (1995)
10. A. Bradley, J. Yao, J. Dewald, C.-P. Richter, Evaluation of electroencephalography source localization algorithms with multiple cortical sources. PLoS ONE **11**(1), e0147266 (2016)
11. M. Breakspear, Dynamic models of large-scale brain activity. Nat. Neurosci. **20**(3), 340 (2017)
12. M.J. Brookes, C.M. Stevenson, G.R. Barnes et al., Beamformer reconstruction of correlated sources using a modified source model. Neuroimage **34**(4), 1454–1465 (2007)
13. R.B. Buxton, E.C. Wong, L.R. Frank, Dynamics of blood flow and oxygenation changes during brain activation: the balloon model. Magn. Reson. Med. **39**(6), 855–864 (1998)
14. R.B. Buxton, K. Uludağ, D.J. Dubowitz, T.T. Liu, Modeling the hemodynamic response to brain activation. Neuroimage **23**, S220–S233 (2004)
15. V.D. Calhoun, T. Adali, G. Pearlson, K. Kiehl, Neuronal chronometry of target detection: fusion of hemodynamic and event-related potential data. Neuroimage **30**(2), 544–553 (2006)
16. D. Cohen, B.N. Cuffin, Demonstration of useful differences between magnetoencephalogram and electroencephalogram. Electroencephalogr. Clin. Neurophysiol. **56**(1), 38–51 (1983)
17. A.M. Dale, M.I. Sereno, Improved localizadon of cortical activity by combining EEG and MEG with MRI cortical surface reconstruction: a linear approach. J. Cogn. Neurosci. **5**(2), 162–176 (1993)
18. A.M. Dale, B. Fischl, M.I. Sereno, Cortical surface-based analysis: I. Segmentation and surface reconstruction. Neuroimage **9**(2), 179–194 (1999)
19. A.M. Dale, A.K. Liu, B.R. Fischl et al., Dynamic statistical parametric mapping: combining fMRI and MEG for high-resolution imaging of cortical activity. Neuron **26**(1), 55–67 (2000)
20. J. Daunizeau, C. Grova, J. Mattout et al., Assessing the relevance of fMRI-based prior in the EEG inverse problem: a Bayesian model comparison approach. IEEE Trans. Signal Process. **53**(9), 3461–3472 (2005)
21. J. Daunizeau, C. Grova, G. Marrelec et al., Symmetrical event-related EEG/fMRI information fusion in a variational Bayesian framework. Neuroimage **36**(1), 69–87 (2007)
22. B.G. Edwards, V.D. Calhoun, K.A. Kiehl, Joint ICA of ERP and fMRI during error-monitoring. Neuroimage **59**(2), 1896–1903 (2012)
23. S. Fazli, J. Mehnert, J. Steinbrink et al., Enhanced performance by a hybrid NIRS–EEG brain computer interface. Neuroimage **59**(1), 519–529 (2012)
24. B. Fischl, A. Liu, A.M. Dale, Automated manifold surgery: constructing geometrically accurate and topologically correct models of the human cerebral cortex. IEEE Trans. Med. Imaging **20**(1), 70–80 (2001)

25. B. Fischl, D.H. Salat, E. Busa et al., Whole brain segmentation: automated labeling of neuroanatomical structures in the human brain. Neuron **33**(3), 341–355 (2002)
26. J.M. Ford, B.J. Roach, V.A. Palzes, D.H. Mathalon, Using concurrent EEG and fMRI to probe the state of the brain in schizophrenia. NeuroImage-Clin. **12**, 429–441 (2016)
27. K. Friston, L. Harrison, J. Daunizeau et al., Multiple sparse priors for the M/EEG inverse problem. Neuroimage **39**(3), 1104–1120 (2008)
28. K. Friston, C. Chu, J. Mourão-Miranda et al., Bayesian decoding of brain images. Neuroimage **39**(1), 181–205 (2008)
29. K.J. Friston, W. Penny, C. Phillips et al., Classical and Bayesian inference in neuroimaging: theory. Neuroimage **16**(2), 465–483 (2002)
30. M. Fuchs, M. Wagner, J. Kastner, Boundary element method volume conductor models for EEG source reconstruction. Clin. Neurophysiol. **112**(8), 1400–1407 (2001)
31. L. Geddes, L. Baker, The specific resistance of biological material—a compendium of data for the biomedical engineer and physiologist. Med. Biol. Eng. **5**(3), 271–293 (1967)
32. G.H. Golub, M. Heath, G. Wahba, Generalized cross-validation as a method for choosing a good ridge parameter. Technometrics **21**(2), 215–223 (1979)
33. S.I. Gonçalves, J.C. de Munck, J.P. Verbunt et al., In vivo measurement of the brain and skull resistivities using an EIT-based method and realistic models for the head. IEEE Trans. Biomed. Eng. **50**(6), 754–767 (2003)
34. R.L. Goris, T. Putzeys, J. Wagemans, F.A. Wichmann, A neural population model for visual pattern detection. Psychol. Rev. **120**(3), 472 (2013)
35. R. Grech, T. Cassar, J. Muscat et al., Review on solving the inverse problem in EEG source analysis. J. NeuroEng. Rehabil. **5**(1), 25 (2008)
36. D. Gutiérrez, A. Nehorai, C.H. Muravchik, Estimating brain conductivities and dipole source signals with EEG arrays. IEEE Trans. Biomed. Eng. **51**(12), 2113–2122 (2004)
37. M. Hämäläinen, R. Hari, R.J. Ilmoniemi et al., Magnetoencephalography—theory, instrumentation, and applications to noninvasive studies of the working human brain. Rev. Mod. Phys. **65**(2), 413 (1993)
38. H. Hallez, B. Vanrumste, R. Grech et al., Review on solving the forward problem in EEG source analysis. J. NeuroEng. Rehabil. **4**(1), 46 (2007)
39. M.S. Hamalainen, *Interpreting Measured Magnetic Fields of the Brain: Estimates of Current Distributions* (Helsinki University of Technology, Rep, 1984)
40. Hansen PC (1999) The L-curve and its use in the numerical treatment of inverse problems
41. O. Hauk, D.G. Wakeman, R. Henson, Comparison of noise-normalized minimum norm estimates for MEG analysis using multiple resolution metrics. Neuroimage **54**(3), 1966–1974 (2011)
42. R.J. Huster, S. Debener, T. Eichele, C.S. Herrmann, Methods for simultaneous EEG-fMRI: an introductory review. J. Neurosci. **32**(18), 6053–6060 (2012)
43. B.H. Jansen, V.G. Rit, Electroencephalogram and visual evoked potential generation in a mathematical model of coupled cortical columns. Biol. Cybern. **73**(4), 357–366 (1995)
44. J. Jorge, W. Van der Zwaag, P. Figueiredo, EEG–fMRI integration for the study of human brain function. Neuroimage **102**, 24–34 (2014)
45. Y. Lai, W. Van Drongelen, L. Ding et al., Estimation of in vivo human brain-to-skull conductivity ratio from simultaneous extra-and intra-cranial electrical potential recordings. Clin. Neurophysiol. **116**(2), 456–465 (2005)
46. X. Lei, C. Qiu, P. Xu, D. Yao, A parallel framework for simultaneous EEG/fMRI analysis: methodology and simulation. Neuroimage **52**(3), 1123–1134 (2010)
47. X. Lei, D. Ostwald, J. Hu et al., Multimodal functional network connectivity: an EEG-fMRI fusion in network space. PLoS ONE **6**(9), e24642 (2011)
48. R. Li, T. Potter, W. Huang, Y. Zhang, Enhancing performance of a hybrid EEG-fNIRS system using channel selection and early temporal features. Front. Hum. Neurosci. **11**, 462 (2017)
49. A.K. Liu, J.W. Belliveau, A.M. Dale, Spatiotemporal imaging of human brain activity using functional MRI constrained magnetoencephalography data: Monte Carlo simulations. Proc. Natl. Acad. Sci. **95**(15), 8945–8950 (1998)

50. Y. Liu, Y. Ning, S. Li et al., Three-dimensional innervation zone imaging from multi-channel surface EMG recordings. Int. J. Neural Syst. **25**(06), 1550024 (2015)
51. Z. Liu, F. Kecman, B. He, Effects of fMRI–EEG mismatches in cortical current density estimation integrating fMRI and EEG: a simulation study. Clin. Neurophysiol. **117**(7), 1610–1622 (2006)
52. Z. Liu, B. He, fMRI–EEG integrated cortical source imaging by use of time-variant spatial constraints. Neuroimage **39**(3), 1198–1214 (2008)
53. M. Luessi, S.D. Babacan, R. Molina et al., Bayesian symmetrical EEG/fMRI fusion with spatially adaptive priors. Neuroimage **55**(1), 113–132 (2011)
54. J. Malmivuo, R. Plonsey, *Bioelectromagnetism: Principles and Applications of Bioelectric and Biomagnetic Fields* (Oxford University Press, USA, 1995)
55. E. Martınez-Montes, P.A. Valdés-Sosa, F. Miwakeichi et al., Concurrent EEG/fMRI analysis by multiway partial least squares. NeuroImage **22**(3), 1023–1034 (2004)
56. T. Medani, D. Lautru, D. Schwartz et al., FEM method for the EEG forward problem and improvement based on modification of the saint venant's method. Prog Electromagn Res **153**, 11–22 (2015)
57. J.W. Meijs, O.W. Weier, M.J. Peters, A. Van Oosterom, On the numerical accuracy of the boundary element method (EEG application). IEEE Trans. Biomed. Eng. **36**(10), 1038–1049 (1989)
58. M.M. Monti, Statistical analysis of fMRI time-series: a critical review of the GLM approach. Front. Hum. Neurosci. **5**, 28 (2011)
59. M. Moosmann, T. Eichele, H. Nordby et al., Joint independent component analysis for simultaneous EEG–fMRI: principle and simulation. Int. J. Psychophysiol. **67**(3), 212–221 (2008)
60. T. Nguyen, T. Potter, T. Nguyen, et al., EEG source imaging guided by spatiotemporal specific fMRI: toward an understanding of dynamic cognitive processes. Neural Plast. (2016)
61. T. Nguyen, T. Potter, R. Grossman, Y. Zhang, Characterization of dynamic changes of current source localization based on spatiotemporal fMRI constrained EEG source imaging. J. Neural Eng. (2017)
62. T.F. Oostendorp, J. Delbeke, D.F. Stegeman, The conductivity of the human skull: results of in vivo and in vitro measurements. IEEE Trans. Biomed. Eng. **47**(11), 1487–1492 (2000)
63. D. Ostwald, C. Porcaro, A.P. Bagshaw, An information theoretic approach to EEG–fMRI integration of visually evoked responses. Neuroimage **49**(1), 498–516 (2010)
64. R.D. Pascual-Marqui, C.M. Michel, D. Lehmann, Low resolution electromagnetic tomography: a new method for localizing electrical activity in the brain. Int. J. Psychophysiol. **18**(1), 49–65 (1994)
65. R.D. Pascual-Marqui, Review of methods for solving the EEG inverse problem. Int. J. Bioeletromagn. **1**(1), 75–86 (1999)
66. R.D. Pascual-Marqui, Standardized low-resolution brain electromagnetic tomography (sLORETA): technical details. Methods Find. Exp. Clin. Pharmacol. **24**(Suppl D), 5–12 (2002)
67. S. Peca, C.R. McCreary, E. Donaldson et al., Neurovascular decoupling is associated with severity of cerebral amyloid angiopathy. Neurology **81**(19), 1659–1665 (2013)
68. Y. Peng, J. He, B. Yao et al., Motor unit number estimation based on high-density surface electromyography decomposition. Clin. Neurophysiol. **127**(9), 3059–3065 (2016)
69. T.K. Perrachione, S.S. Ghosh, Optimized design and analysis of sparse-sampling FMRI experiments. Front. Neurosci. **7**, 55 (2013)
70. S.E. Robinson, J. Vrba, *Functional Neuroimaging by Synthetic Aperture Magnetometry (SAM). Recent Advances in Biomagnetism* (Tokyo University Press, Sendai, Japan, 1999), pp 302–305
71. M. Rosa, J. Daunizeau, K.J. Friston, EEG-fMRI integration: a critical review of biophysical modeling and data analysis approaches. J. Integr. Neurosci. **9**(04), 453–476 (2010)
72. S. Rush, D.A. Driscoll, Current distribution in the brain from surface electrodes. Anesth. Analg. **47**(6), 717–723 (1968)
73. S. Rush, D.A. Driscoll, EEG electrode sensitivity-an application of reciprocity. IEEE Trans. Biomed. Eng. **1**, 15–22 (1969)

74. Y. Salu, L.G. Cohen, D. Rose et al., An improved method for localizing electric brain dipoles. IEEE Trans. Biomed. Eng. **37**(7), 699–705 (1990)
75. U. Schridde, M. Khubchandani, J.E. Motelow et al., Negative BOLD with large increases in neuronal activity. Cereb. Cortex **18**(8), 1814–1827 (2007)
76. K. Sekihara, S. Nagarajan, D. Poeppel, Y. Miyashita, Reconstructing spatio-temporal activities of neural sources from magnetoencephalographic data using a vector beamformer, in *Acoustics, Speech, and Signal Processing, 2001 Proceedings (ICASSP'01) 2001 IEEE International Conference on 2021–2024* (2001)
77. D. Sharon, M.S. Hämäläinen, R.B. Tootell et al., The advantage of combining MEG and EEG: comparison to fMRI in focally stimulated visual cortex. Neuroimage **36**(4), 1225–1235 (2007)
78. K. Uutela, M. Hamalainen, R. Salmelin, Global optimization in the localization of neuromagnetic sources. IEEE Trans. Biomed. Eng. **45**(6), 716–723 (1998)
79. B.D. Van Veen, K.M. Buckley, Beamforming: a versatile approach to spatial filtering. IEEE ASSP Mag. **5**(2), 4–24 (1988)
80. B.D. Van Veen, W. Van Drongelen, M. Yuchtman, A. Suzuki, Localization of brain electrical activity via linearly constrained minimum variance spatial filtering. IEEE Trans. Biomed. Eng. **44**(9), 867–880 (1997)
81. Y. Zhang, S. Zhu, L. Ding, et al., On the forward problem of EEG cortical imaging by means of finite element method. Engineering in Medicine and Biology Society, in *IEMBS'04 26th Annual International Conference of the IEEE*, pp 440–442 (2004)
82. Y. Zhang, S. Zhu, B. He, A second-order finite element algorithm for solving the three-dimensional EEG forward problem. Phys. Med. Biol. **49**(13), 2975 (2004)
83. Y. Zhang, W. Van Drongelen, B. He, Estimation of in vivo brain-to-skull conductivity ratio in humans. Appl. Phys. Lett. **89**(22), 223903 (2006)
84. Y. Zhang, L. Ding, W. van Drongelen et al., A cortical potential imaging study from simultaneous extra-and intracranial electrical recordings by means of the finite element method. Neuroimage **31**(4), 1513–1524 (2006)
85. Y. Zhang, W. van Drongelen, M. Kohrman, B. He, Three-dimensional brain current source reconstruction from intra-cranial ECoG recordings. Neuroimage **42**(2), 683–695 (2008)
86. Y. Zhang, D. Wang, G.W. Timm, A three-dimensional muscle activity imaging technique for assessing pelvic muscle function. Inverse Probl **26**(11), 115018 (2010)
87. Zhang Y (2013) Noninvasive imaging of internal muscle activities from multi-channel surface EMG recordings, in *Engineering in Medicine and Biology Society (EMBC), 2013 35th Annual International Conference of the IEEE*, pp 5430–5432

Chapter 6
Methods for Functional Connectivity Analysis

Jeong Woo Choi and Kyung Hwan Kim

Abstract The purpose of this chapter is to provide comprehensive and useful guidelines for the methods of the functional connectivity analysis (FCA) for electroencephalogram (EEG) and its application. After presenting the detailed procedure for the FCA, we described various methods for quantifying functional connectivity. The problem of volume conduction and the means to diminish its confounding effects on the FCA was thoroughly reviewed. As a useful preprocessing for the FCA, spatial filtering of the time-series measured on the scalp or transformation to current densities on cortical surface were described. We also reviewed ongoing efforts toward developing FC measures which are inherently robust to the volume conduction problem. Finally, we illustrated the procedures for determining significance of the FC among specific pair of regions, which exploit surrogate data generation or the characteristics of event-related data.

6.1 Introduction

Cognition and behavior is enabled by coordinated and integrated activities of neuronal populations of relevant regions in the brain. Beyond spatial and temporal pattern of brain activation, investigating the interaction between those neuronal populations, i.e., the functional connectivity analysis (FCA), is essential for proper understanding of human brain function [6, 17, 49, 56]. Now the FCA is regarded as one of the major tools for functional brain imaging.

In functional neuroimaging studies, mainly using functional magnetic resonance imaging (fMRI), intrinsic cortical networks such as default mode and saliency networks, have been identified during both resting state and task performance by the FCA [7, 17, 55]. The functional brain network is obviously dynamic although most fMRI-based FCA studies so far implicitly assumed static functional connectivity

---

J. W. Choi · K. H. Kim (✉)
Department of Biomedical Engineering, Yonsei University, Wonju, South Korea
e-mail: khkim0604@yonsei.ac.kr

C.-H. Im (ed.), *Computational EEG Analysis*, Biological and Medical Physics, Biomedical Engineering, https://doi.org/10.1007/978-981-13-0908-3_6

pattern. Dynamic FCA of fMRI blood oxygen level-dependent (BOLD) signals is currently under active investigation [26].

Considering the intrinsic limitation in temporal resolution of fMRI, electrophysiological recordings of neural activity are better suited for the dynamic FCA, especially for the investigation of short-term neural phenomena with temporal resolution of millisecond scale. Either invasive or noninvasive recording techniques can be used for the FCA. However, noninvasive methods, i.e. electroencephalogram (EEG) and magnetoencephalogram (MEG), are to be used for human behavioral/cognitive neuroscience studies under experimental task or task-free resting state.

The EEG/MEG signals are obtained from an array of sensors placed on the scalp, so the spread of electromagnetic fields prohibits direct interpretation of spatial origin of the signals from a single channel. Localization of cortical current sources is obtained by solving an electromagnetic inverse problem [4, 33, 41], and it may be applied prior to the FCA to investigate the connectivity between specific brain regions [20, 46]. This may be even crucial for valid interpretation of the FCA results in that functional connection between specific cortical regions can be identified. Source imaging techniques using distributed source models are being combined with various measures of the FC, providing significant results on cognitive, behavioral, and clinical results [1, 2, 11, 25, 29, 39]. The high temporal resolution of EEG/MEG can also be utilized to investigate coupling between different rhythms in various frequency bands present within neural activities.

The FC measures should reflect the association of neural activities in different brain regions. Hence, they should quantify the correlation and/or causality between the time-series of neural activities of multiple brain areas [6, 15, 42, 47]. Linear correlation coefficient is still one of the most commonly used measure of the FCA for fMRI. Various measures have origins from various disciplines such as statistical signal processing, nonlinear dynamics, and information theory, and they have been adopted for the FCA analysis in order to deal with complicated interaction between neuronal populations [6, 15, 42, 47].

It is recognized that oscillatory neural activities represent formation of local neuronal populations [9], and underlie dynamic coordination of brain function and synaptic plasticity [6, 50, 60]. Therefore, the interaction between oscillatory rhythmic activities should provide valuable insights on inter-regional communication among neuronal population, and MEG and EEG are the most suitable for this purpose. Novel measures for better analysis of the couplings between rhythmic activities are under active research and being applied for the FCA of EEG/MEG [5, 20, 24], exploiting the high resolution of these electrophysiological signals. Beyond coupling of rhythms within a single frequency band, cross-frequency couplings have been explored by quantifying either phase-phase or phase-amplitude couplings [10, 45, 59].

The purpose of this review article is to provide comprehensive and useful guidelines on the methods and to illustrate application of the FCA for EEG. Although the target is on EEG, the contents may be useful for the FCA of MEG as well. The focus is on how the FCA can be properly applied to cognitive neuroscience studies

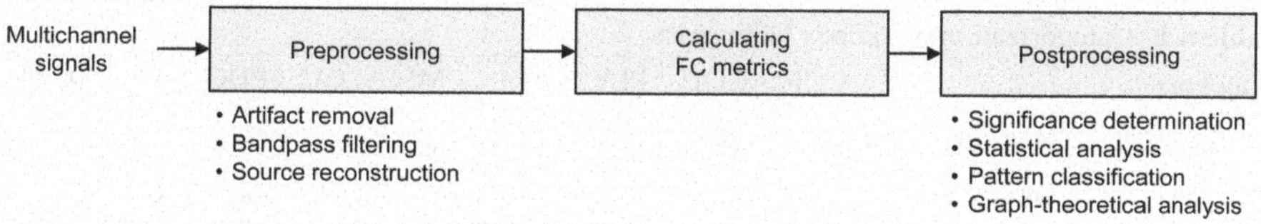

Fig. 6.1 Illustration of the procedure for the FCA of multichannel EEGs

and clinical investigations. The merits and pitfalls of each FC measures are to be illustrated so that the readers may find this article useful to select the best method among many options available.

6.2 Procedure for the FCA

Figure 6.1 illustrates the detailed procedure for the FCA of EEG. Multichannel signals are preprocessed, primarily for the removal of artifacts including eye blinks and movements, muscle activity, and skin potentials. Bandpass filtering is often applied to extract the oscillatory rhythms within the frequency bands of interest. For the FCA between cortical regions, the time-series in the sensor space are projected onto the cortical source space using distributed source imaging techniques [4, 33, 40]. The multiple time-series are then subject to the calculation of FC measures between channels or cortical regions, which yields a functional connectivity matrix. Each element of the matrix quantifies the connectivity between two specific regions.

Sometimes the elements of the FC matrix are transformed to either 1 or 0 by determining the significant and insignificant connections by comparing the threshold level determined by surrogate data [15, 30, 54]. Then statistical comparisons are applied in order to determine the significant differences among experimental conditions or subject groups. Multivariate pattern analysis based on machine-learning can also be applied so that the information regarding conditions or groups can be decoded from the connectivity matrix [31, 35]. The adjacency matrix can be regarded as a graph with nodes and edges [8, 53], and thus, the pattern of the connectivity can be further characterized by graph theory [13, 14, 51, 57].

6.3 FC Metrics

There are many FC metrics with different theoretical backgrounds such as statistical signal processing, time-series forecasting, information theory, and nonlinear dynamics [6, 15, 42, 44]. It is often unclear which method should be used. They can be categorized by their features including theoretical basis, directionality, and signal domains. Table 6.1 summarizes various FC metrics to be described in depth in this review article, in terms of these features.

Table 6.1 Categorization of various FC metrics

Category		CCF	COH	PLV	PLI	MI	GC	PDC	TE	DCM
Directionality	Non-directed	√	√	√	√	√				
	Directed						√	√	√	√
Theoretical basis	Data-driven	√	√	√	√		√	√		
	Information-based					√			√	
	Model-based									√
Signal domain	Amplitude	√	√			√	√	√	√	√
	Phase		√	√	√					

FC functional connectivity, *CCF* Cross-correlation function, *COH* Coherence, *PLV* phase locking value, *PLI* phase lag index, *MI* mutual information, *GC* Granger's causality, *PDC* partial directed coherence, *TE* transfer entropy, *DCM* dynamic causal modeling

6.3.1 *Cross-Correlation Function (CCF)*

CCF is defined as the linear correlation between two signals represented as a function of the time delay between them. The CCF between two signals, $x(t)$ and $y(t)$, is calculated as follows:

$$CCF_{x,y}(\tau) = \frac{1}{N-\tau}\sum_{t=1}^{N-\tau}\left(\frac{x(t+\tau)-\bar{x}}{\sigma_x}\right)\left(\frac{y(t)-\bar{y}}{\sigma_y}\right). \tag{6.1}$$

Here, N is the total number of samples of the signals, and τ is the time delay between the two signals. $\bar{x}$ and σ_x denote mean and standard deviation of the signal x, respectively. The CCF ranges between -1 (perfect inverse correlation) and 1 (perfect correlation), and equals zero for the case of no correlation at the time delay τ. The CCF at time delay of 0 is the Pearson's correlation coefficient.

6.3.2 *Coherence*

The coherence represents the linear correlation between two signals x and y calculated in the frequency domain, which is calculated as follows:

$$COH_{x,y}(f) = \frac{\left|\left\langle S_{x,y}(f)\right\rangle\right|}{\sqrt{\left\langle S_{x,x}(f)\right\rangle \cdot \left\langle S_{y,y}(f)\right\rangle}}. \tag{6.2}$$

Here $\langle\cdot\rangle$ indicates average over a predetermined time interval. $S_{x,y}$ represents the cross spectral density function (CSDF) of two signals, x and y, which is derived by Fourier transform of the CCF in Sect. 6.3.1. The definition of coherence includes the normalization of CSDF $S_{x,y}$ by auto-spectral density functions, $S_{x,x}$ and $S_{y,y}$, so that the range of coherence becomes between 0 and 1.

It should be noted that the coherence is still sensitive to spectral power even though its definition contains the normalization by spectral powers of the two signals. Thus, it is often unclear whether the coherence at a specific frequency is dominated by powers of the signals and/or phase relationships between them [30].

6.3.3 *Phase Locking Value (PLV)*

PLV measures the degree of phase locking between two signals over time, by observing whether the phase difference between them is relatively constant within a temporal interval. Prior to calculating the PLV, the signals are first transformed into narrowband signal in the frequency band of interest (e.g., theta or gamma band) by bandpass filtering. The instantaneous phase angle, $\phi(t)$ is calculated from the narrowband signal $x(t)$ and its Hilbert transform, $\tilde{x}(t)$, as follows [23, 30]:

$$\phi(t) = \arctan \frac{\tilde{x}(t)}{x(t)}.$$

The PLV between two signals x and y is calculated by averaging the phase difference over N time points as follows [30]:

$$PLVx, y = \frac{1}{N}\left|\sum_{t=1}^{N} \exp[i\{\phi_x(t) - \phi_y(t)\}]\right|. \tag{6.3}$$

Here, $\phi_x(t)$ and $\phi_y(t)$ represent the instantaneous phase angles for each time point t for two signals, x and y, respectively. PLV ranges between 0 (no synchronization) and 1 (perfect synchronization).

6.3.4 *Phase Lag Index (PLI)*

The PLI was developed to mitigate the spurious phase synchrony resulting from common sources, due to volume conduction or active reference electrodes [52]. This will be described in detail later in Sect. 6.4. The PLI is defined to quantify the asymmetry of the distribution of phase differences between two signals (i.e. either positive or negative phase differences). This asymmetry implies the presence of nonzero phase difference (i.e., time lag) between two signals. If the phase synchrony is due to the common sources, the phase differences are expected to be symmetrically distributed around zero.

The calculation of PLI is similar to that of the PLV, and involves bandpass filtering and Hilbert transform as follows [52]:

$$PLI_{x,y} = \frac{1}{N}\left|\sum_{t=1}^{N} sign(\phi_x(t) - \phi_y(t))\right|. \tag{6.4}$$

here *sign* represents the sign of the phase difference (i.e., −1 for negative, 1 for positive, and 0 for zero values, respectively). The PLI ranges between 0 (no synchronization) and 1 (perfect synchronization).

6.3.5 Mutual Information (MI)

MI quantifies the amount of information that two signals share each other based on a basic measure of information, *Shannon entropy* [48]. *Shannon entropy* is defined as the average amount of information (or code) which is necessary to encode a discrete variable [42, 48]. The entropy $H(X)$ is calculated as follows:

$$H(X) = -\sum_{i=1}^{n} p(x_i)\log_2 p(x_i). \tag{6.5}$$

Here, $p(x_i)$ is the probability of the values of the signal x in the ith bin, and n represents the number of bins used to construct a histogram which approximates the probability density function (PDF) of x. The entropy is positive and has a unit in bits, and unrelated to the temporal structure of the signal. It is important to estimate the appropriate number of bins, since the approximation of PDF by a histogram is sensitive to the number of bins [15]. Diaconis and Freedman [16] suggested a guideline for an optimal number of bins as follows [16]:

$$nbins = \left\lceil \frac{\max(x) - \min(x)}{2Q_x n^{-1/3}} \right\rceil, \tag{6.6}$$

where Q_x is the range between the 25th and the 75th percentiles of data distribution X, n represents the total number of data points, and $\max(x)$ and $\min(x)$ are the maximum and minimum values of x, respectively.

From the entropies of the two signals x and y, and their joint entropy, i.e., $H(X)$, $H(y)$, and $H(X,Y)$, MI is calculated as follows:

$$MI_{x,y} = H(X) + H(Y) - H(X,Y). \tag{6.7}$$

Also, $H(X,Y)$ is the joint entropy between two signals, and defined as follows:

$$H(X, Y) = -\sum_{j=1}^{m}\sum_{i=1}^{n} p(x_i, y_j)\log_2 p(x_i, y_j), \tag{6.8}$$

where $p(x_i, y_j)$ is the joint probability of the values of the signal x in the *ith* bin and the signal y in the *jth* bin. If there is no relationship between two signals at all, X and Y are independent, and thus, the joint probability $p(x_i, y_j)$ is equivalent to $p(x_i)p(y_j)$. Hence, the joint entropy $H(X, Y)$ will be $H(X) + H(Y)$, and the MI becomes zero. Otherwise, the MI should be positive and would show the maximum value when two signals are equal.

6.3.6 Granger Causality (GC)

The idea of GC is that signal x causes signal y if the prediction error of y estimated by autoregressive (AR) modeling is significantly reduced when it is estimated by joint AR modeling of x and y [19]. This can be assessed by comparing the univariate and bivariate AR models for the two signals, x and y.

The univariate AR models for each signal, x and y, are described as follows [42]:

$$x(t) = \sum_{n=1}^{p} a_{x,n}x(t-n)+e_x(t), \quad y(t) = \sum_{n=1}^{p} a_{y,n}y(t-n)+e_y(t). \tag{6.9}$$

Here, p denotes the number of lagged observations included in the model (i.e., model order), and $a_{x,n}$ and $a_{y,n}$ are the model coefficients at time lag n, and e_x and e_y are the prediction error for each signal estimated by the model. The prediction error depends on the past values of the signal.

Alternatively, the joint, bivariate AR model of x and y is as follows:

$$\begin{aligned} x(t) &= \sum_{n=1}^{p} a_{x,y,n}x(t-n)+\sum_{n=1}^{p} b_{x,y,n}y(t-n)+e_{x,y}(t) \\ y(t) &= \sum_{n=1}^{p} a_{y,x,n}y(t-n)+\sum_{n=1}^{p} b_{y,x,n}x(t-n)+e_{y,x}(t) \end{aligned}. \tag{6.10}$$

Here, p is the model order, a and b contain the coefficients of the model, and $e_{x,y}$ and $e_{y,x}$ denote the prediction errors of the signals estimated by the model. Here the prediction error depends on the past values of both signals.

The prediction performances of the univariate and bivariate models can be compared quantitatively from the variances of the prediction errors as follows:

$$V_{x|x} = \text{var}(e_x) \text{ and } V_{y|y} = \text{var}(e_y) \text{ for univariate AR model,}$$

$V_{x|x,y} = \text{var}(e_{x,y})$ and $V_{y|y,x} = \text{var}(e_{y,x})$ for bivariate AR model.

Here, var(·) denotes the variance.

The Granger causality between two signals, x and y, is calculated as the log-ratio of variances follows:

$$GC_{x\to y} = \ln\left(\frac{V_{y|y}}{V_{y|y,x}}\right) \text{ for the measure of 'signal x causes signal y'}$$

$$GC_{y\to x} = \ln\left(\frac{V_{x|x}}{V_{x|x,y}}\right) \text{ for the measure of 'signal y causes signal x'}$$

The prediction error of y should not be reduced whether x is considered or not for the estimation of y, if there exist no causal influence from x to y. This implies that the variances $V_{y|y}$ and $V_{y|y,x}$ are identical, and thus, $GC_{x\to y}$ is close to zero. On the other hand, causal influence of x to y reduces the prediction error of y when x is considered. Hence, $GC_{x\to y}$ becomes a positive value. The GC measure is directional. If the GCs of both directions are high, it can be interpreted as a bidirectional connectivity [42].

6.3.7 Partial Directed Coherence (PDC)

PDC is a frequency domain equivalent of the GC, based on multivariate autoregressive (MVAR) modeling of multichannel signals [3]. Let's assume that the simultaneously recorded m channel signals $\mathbf{x}(t) = [x_1(t), \ldots, x_m(t)]^T$ can be described by an MVAR model as follows:

$$\mathbf{x}(t) = \sum_{n=1}^{p} \mathbf{A}_n \mathbf{x}(t-n) + \mathbf{e}(t). \tag{6.11}$$

here, p is the model order, $\mathbf{A}_n = \begin{bmatrix} a_{1,1}(n) & \cdots & a_{1,m}(n) \\ \vdots & \ddots & \vdots \\ a_{m,1}(n) & \cdots & a_{m,m}(n) \end{bmatrix}$ is the matrix of model coefficients at time lag n, and $\mathbf{e}(t) = [e_1(t), \ldots, e_m(t)]^T$ is a multivariate Gaussian white noise with zero mean and covariance matrix $\boldsymbol{\Sigma}$. The model coefficients $a_{m,m}$ indicate the influence among the signals (e.g., $a_{1,2}(n)$ is the influence of $x_2(t-n)$ on $x_1(t)$).

This time domain representation can be transformed into frequency domain by Fourier transform (FT). $\bar{\mathbf{A}}(f) = \mathbf{I} - \mathbf{A}(f) = [\bar{\mathbf{a}}_1(f)\bar{\mathbf{a}}_2(f)\ldots\bar{\mathbf{a}}_m]$, where $\mathbf{A}(f)$ is the FT of the model coefficients and $\bar{a}_{i,j}(f)$ is the i,jth element of $\bar{\mathbf{A}}(f)$. The PDC from signal x_i to signal x_j can be calculated as follows:

$$PDC_{i\to j}(f) = \frac{\bar{a}_{i,j}(f)}{\sqrt{\bar{\mathbf{a}}_j^H(f)\bar{\mathbf{a}}_j(f)}}, \tag{6.12}$$

where H indicates the transpose and complex conjugate operator. Thus, the PDC quantifies relative strength of the influence of the signal x_i on the signal x_j at frequency f.

Another metric based on a MVAR model, directed transfer function (DTF), was proposed [27]. The DTF is quite similar to the PDC metric in that it reveals causal relations between time-series based on a MVAR model. However, the DTF can be calculated from the transfer function matrix, H, instead of A for the PDC calculation, where those two matrices are related as $H(f) = \bar{A}^{-1}(f)$. Because of the matrix inversion, DTF demands higher computational loads and may suffer from numerical imprecisions due to potential ill-conditioning of $\bar{A}(f)$ [3]. If the structure of the matrix $H(f)$ is preserved upon inversion, the DTF and PDC lead to identical results for the effective connectivity [3].

6.4 Volume Conduction Problem

As explained above, there exist numerous methods for the FCA of EEG (and MEG), which originated from various theoretical backgrounds. Considering the possibility of combining various preprocessing, FC measure, and postprocessing methods available, the choice of appropriate FCA method is far from obvious in most applications since each method has its own pros and cons, rendering the interpretation of the results ambiguous.

There exist several issues that deserve caution when interpreting the FCA results. For example, the estimated FC may reflect the true neuronal interaction or not. This is related to the fact that EEG (MEG as well) signals include both relevant and irrelevant signals and/or noises. Moreover, it is not possible to make sure whether the observed connectivity is due to direct or indirect one through an unobserved pathway. Besides, common reference problem and low signal-to-ratio causes significant amount of errors. In addition to the noise or artifact, the FCA results may be affected by the difference of signal-to-noise ratio between channels. Especially this has a huge effect on the estimation of information flow direction. A recent review paper provides a detailed discussion on these issues focusing on oscillatory coupling [6].

Methods have been developed to overcome aforementioned issues. For example, in the case of the FCA based on coupling between rhythmic oscillatory neural activities, the volume conduction problem may be alleviated from the fact that the phases of two rhythmic signals at any pair of locations are different by either 0° or 180°, since the effect of volume conduction and field spread can be regarded as instantaneous [36]. The measures of oscillatory coupling taking this into account have been developed, e.g., PLI [52], imaginary coherence [36], and phase slope index (Nolte et al. [37]). Converting scalp EEGs to current densities on cortical surfaces may be

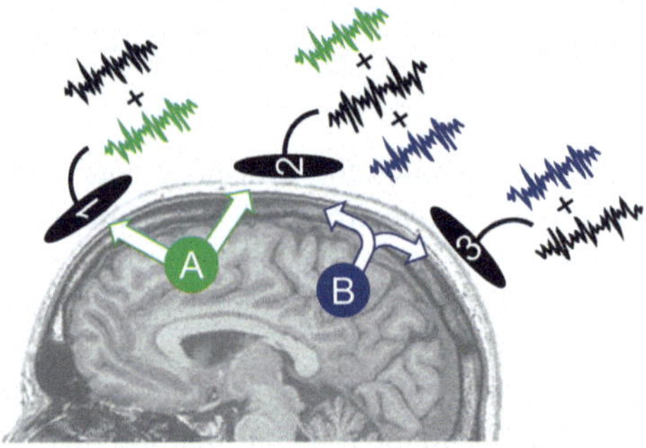

Fig. 6.2 An illustration of the effects of volume conduction on scalp EEGs

greatly helpful since common factors in the signals are significantly reduced. It is investigated which methods for the cortical source localization and FCA provide best results for the FCA [20, 22]. In this section, we try to provide a guideline to reduce the confounding effects of volume conduction in the FCA.

6.4.1 FCA Between the Signals from Surface Electrodes

An EEG electrode placed on scalp surface captures electric potential at a specific location on the scalp. Multiple cortical sources distributed over a wide area on cortical surface contribute to the voltage at a point on the scalp. Conversely, the electric current caused by a localized cortical current source is propagated to a wide area on the scalp. This field spread or volume conduction problem prohibits a rigorous FCA using scalp EEG, and incorrectly emphasizes the functional connections between proximate regions. Figure 6.2 shows examples of volume conduction effect. A single current source (A) affects more than one electrode (1 and 2). Also, the electromagnetic field originating from a single source (B) spreads to multiple adjacent electrodes (2 and 3) through brain tissues such as cerebrospinal fluid, dura, scalp, and skull. These common sources lead to spurious connectivity between scalp EEG channels even though all the cortical current sources are independent [6, 15, 38, 52]. Hence, caution should be made when calculating and interpreting the FC metrics.

Unpredictable phenomena may occur due to the volume conduction effect as illustrated in Fig. 6.3 which is generated from actual 64 channel EEG recordings during an auditory oddball task [12]. First, the phase differences between the EEGs from two nearby electrodes, Fpz and Fp1, were found to be concentrated at zero degree (Fig. 6.3a). Second, it was also found that the strength of connectivity is inversely correlated to the distance between two electrodes (Fig. 6.3b). In particular, the PLV between two closest neighbors showed almost perfect locking (i.e., PLV was close to 1). It was also observed that the connectivity strength is significantly correlated to the spectral power (Fig. 6.3c). Although these are only a few among

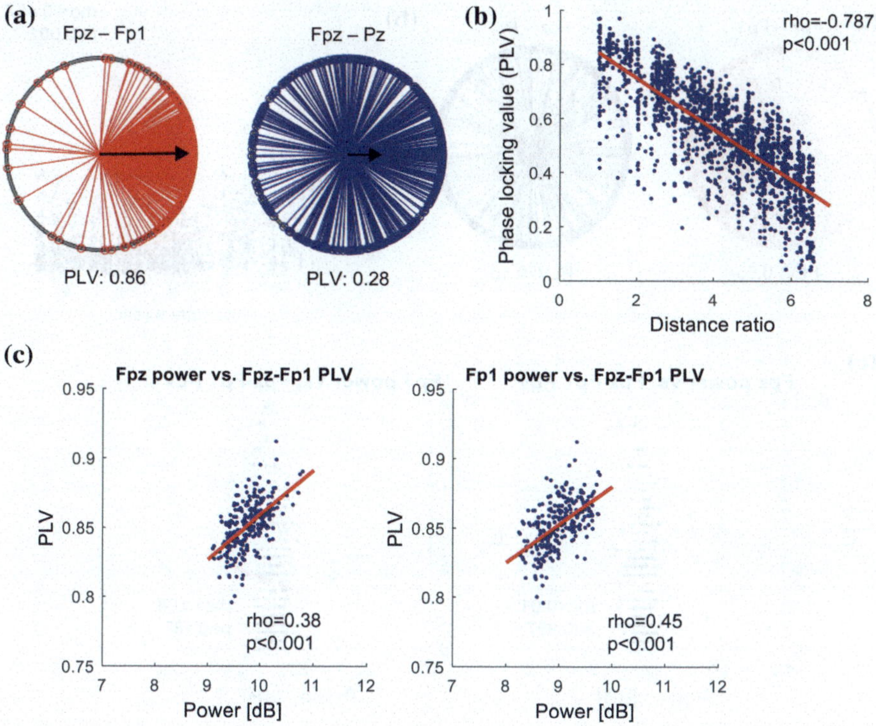

Fig. 6.3 Examples of spurious connectivity (PLV) due to the volume conduction. **a** Distributions of phase differences between two EEG channels in gamma band (30–50 Hz), solid lines denote the phase difference between two channels at a single temporal point, the length of the black arrow corresponds to the magnitude of average PLV. **b** Correlation between PLV and inter-electrode distance in gamma band. **c** Correlation between PLV and spectral power in gamma band

the potential problem of the FC analysis using surface EEG, at least they should be checked to verify whether the conclusions are made by spurious effects of volume conduction.

6.4.2 Spatial Filtering

Surface Laplacian is a method to estimate the amount of current source density (CSD) at the scalp, and behaves as a spatial highpass filter [43]. The potential distribution of EEGs on the scalp usually have low spatial frequency component due to volume conduction. When the surface Laplacian is properly used, the spurious low spatial frequency components may be reduced. As a result, its confounding effect on FC may be eliminated as well (For review of the algorithm of surface Laplacian, see [28, 43]).

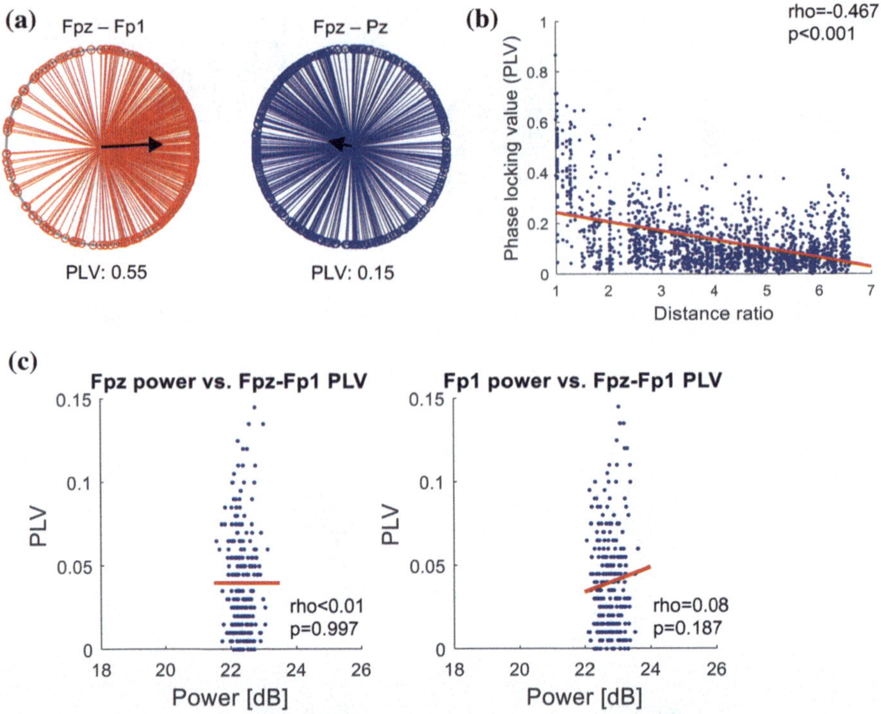

Fig. 6.4 Spatial filtering (surface Laplacian) can reduce the spurious effects of volume conduction. **a** Phase difference between two nearby channels became widely distributed after applying surface Laplacian. The correlation between PLV and inter-electrode distance **b** and between PLV and spectral power **c** became reduced compared to Fig. 6.3

We found that the spurious effects of volume conduction illustrated in Fig. 6.4 were greatly reduced or eliminated by applying surface Laplacian before the FC analysis. After applying the surface Laplacian, the phase differences between two nearby channels (Fpz and Fp1) were much more widely distributed (Fig. 6.4a), in contrast to the previous case of raw EEGs where the phase differences were concentrated around zero degree (Fig. 6.3a). The PLV was decreased to 0.55 after applying surface Laplacian, from 0.86. The correlation between the FC strength and inter-electrode distance was mitigated (from −0.787 to −0.467, Fig. 6.4b). The correlation between the spectral power and the FC in a frequency band became insignificant as well (Fig. 6.4c). In conclusion, the surface Laplacian may provide a partially useful method to reduce the confounding effects of the volume conduction in FC analysis of surface EEG.

6.4.3 FC Measures Robust to Volume Conduction Effect

Another solution is to use the FC metrics which is inherently robust to the volume conduction [34, 36, 38, 52, 58]. Nunez et al. [38] proposed a modified version of coherence, called the reduced coherence. It is calculated by subtracting the random coherence from the measured coherence. Alternatively, partial coherence was introduced, which removes the linear effect of the third time-series (considered as common source) from the coherence calculated from a pair of time-series [34].

Nolte et al. [36] suggested imaginary coherence (ImC). It is based on the hypothesis that the imaginary part of coherence (i.e., the non-zero phase difference) cannot be affected by the volume conduction [36]. In the same vein, Stam et al. [52] proposed the PLI based on how much the non-zero phase differences are distributed to negative (phase lag) or positive (phase lead) sides of the x axis on the complex plane [52]. More recently, an extended version of PLI, called the weighted PLI (WPLI), was suggested to take into account the magnitude as well as the distribution of the phase differences [58].

Figure 6.5 illustrates the advantage of PLI to mitigate the volume conduction effects, as compared to the PLV. The phase differences between EEGs of two nearby channels Fpz and Fp1 are distributed symmetrically around zero degree, which resulted in much lower value of PLI as compared to the PLV (0.09 vs. 0.86). In addition, the correlation between the FC and the inter-electrode distance became drastically reduced to −0.174, from −0.787 in the case of PLV (Fig. 6.5b). The correlation with the spectral power became insignificant as well (Fig. 6.5c). All the results in Fig. 6.5 shows that spurious FC due to the volume connection can be alleviated by using the PLI. Vinck et al. [58] pointed out the problems of the PLI. Temporal discontinuity may occur when there exist small perturbations which lead to phase lags from phase leads (and vice versa) between two time-series. Also, the estimation of PLI is statistically biased, hence its calculation may suffer from the small sample size [52, 58]. Modified versions of the PLI (weighted PLI and debiased weighted PLI) have been proposed to overcome these limitations [58].

6.5 EEG FC Analysis on Source Space

Spatial filters such as surface Laplacian can be applied before the FCA to lessen the effect of volume conduction as shown above. More recently, the reconstruction of cortical current sources is performed prior to the FCA by solving an inverse problem [4, 33, 40]. This provides time-series of cortical current densities at numerous vertices on cortical surface, which enables the calculation of FC measures among cortical regions. The FCA at the cortical source space is advantageous also for the better explanation of the obtained results, since each pair of connection has anatomical interpretation [46].

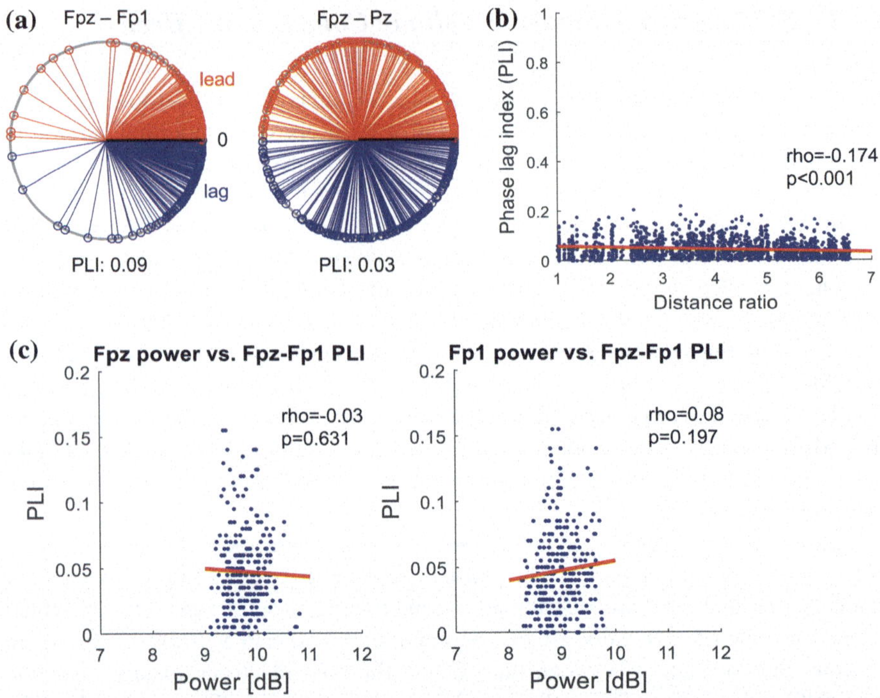

Fig. 6.5 The spurious effects of volume conduction can be reduced by using phase lag index (PLI). **a** Distribution of phase difference between two time-series. The red and blue colors indicate phase lead and phase lag, respectively. **b** Correlation between PLI and inter-electrode distance. **c** Correlation between PLI and spectral power

Figure 6.6 illustrates the detailed procedure of FCA on cortical source space, where solutions of an electromagnetic inverse problem are used to estimate the cortical sources and reconstruct their temporal dynamics. Among several approaches proposed so far, methods based on a distributed cortical source model are appropriate since they aim to provide cortical current time-series at every cortical location (Fig. 6.6b). The most popular ones include the minimum norm estimate (MNE) and its variants (weighted MNE, wMNE), low resolution brain electromagnetic tomography (LORETA), and standardized LORETA (sLORETA). Beamforming methods are also applicable. It is also feasible that the mixed cortical sources due to the volume conduction are 'demixed' by blind source separation [21].

Estimated time-series represent current densities on cortical surface, and they are subject to FC measure calculation. Spatial sampling is commonly used to reduce the number of time-series, or regions of interest (ROIs) are selected before the FCA. The ROI selection is of crucial importance, and based on either a prior knowledge (Fig. 6.6c, image source: http://freesurfer.net) or the results of functional neuroimaging (Fig. 6.6d). Often, the most important ROIs are determined and the cortical maps which represent the crucial regions functionally connected to those ROIs.

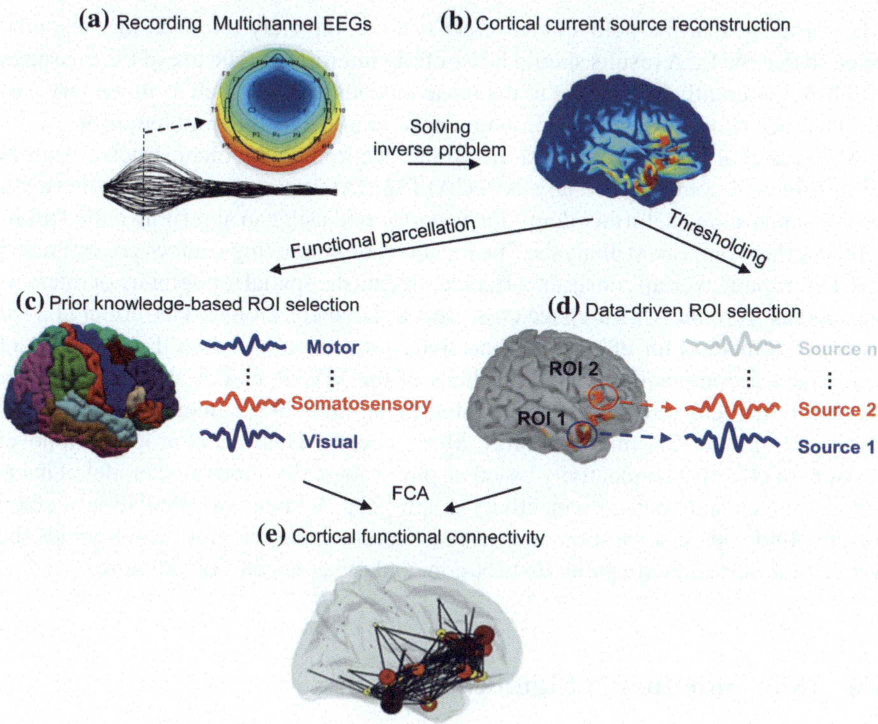

Fig. 6.6 An illustration of procedure of FCA on cortical source space

Hassan et al. [20] reported a comparative study on the processing methods for the FCA on cortical source space [20]. They showed that the results are highly dependent on the selected processing methods as well as the number of electrodes. The combination of wMNE and PLV was found to yield the most relevant result. Their results imply that an optimal combination of source estimation and FCA is essential to correctly identify the functional cortical networks, and thus, the EEG source FCA should be performed carefully in terms of the detailed processing method.

It should be noted that there exist some cases where the spurious result is unavoidable. For example, when the FCA is performed on preselected ROIs, inappropriate ROI selection should lead to incorrect conclusions. The signal-to-noise ratio affects the FCA and may vary systematically according to experimental condition, thereby incorrect significant difference among conditions may be unavoidable.

Aforementioned two-step approaches, consisting of source estimation and FC measure calculation, may yield undesirable incorrect results as has been shown by simulation studies [21], due to several reasons. The unmixing of the scalp EEG signals is far from being perfect regardless of the methods of source estimation. Schoffelen and Gross [46] provides a review of methods for the FCA in cortical source space, focusing on selecting FC measure and region of interests (ROI) [46].

It is warranted that the field spread effect is not completely removed in the source space so that the FCA results should be carefully interpreted. The use of FC measures which are inherently insensitive to the instantaneous mixing, such as imaginary part of coherence (imagcoh), can be recommended to alleviate this problem [46].

Marzetti et al. [32] developed a method to decorrelate the reconstructed sources using principal component analysis (PCA) [32]. Assuming orthogonality between the estimated sources, further demixing is performed using an algorithm called minimum overlap component analysis. The locations of interacting sources are estimated under minimum overlap constraint after identifying the spatial topography of interacting sources from the sensor-space cross-spectral density. Gomez-Herrero et al. [18] presented a method for effective connectivity estimation based on the independent component decomposition of the residuals of the MVAR model, which are probably due to the field spread [18]. The spatial topography of the interacting sources is obtained from the ICA mixing matrix. More recently, Haufe [21] proposed a novel measure of effective connectivity based on physiologically-motivated model of interacting sources and sparse connectivity graph [21]. A one-shot calculation method for the blind source separation and inverse source reconstruction, which yields the source time series, their spatial distribution, and the connectivity structure.

6.6 Determination of Significance

The calculated FC metrics may include false positives due to several confounding effects such as residual artifacts, volume conduction, and common reference. Hence it is important to determine statistical significance. As shown in Fig. 6.7a, null distribution of the FC can be generated from a surrogate data obtained by random shuffling and used to determine significance, which is often defined by the upper 5 or 1% of the null distribution.

Several methods can be used to generate the surrogate data from the experimental data [15, 30, 54]. Random shuffling of the time samples of one of the two time-series destroys the temporal structure. If all time samples are randomly shuffled and the temporal structure is completely destroyed, the null distribution obtained from the surrogate data may result in excessively high false positive rate, i.e., inflate the statistical significance. This can be understood from the fact that the FC measure calculated from any experimental data would be much higher than those calculated from surrogate data, in which the temporal structure is completely destroyed. An alternative is illustrated in Fig. 6.7b, which is called 'time-shift' method [15]. Here one time-series is separated into two segments at a randomly chosen temporal point, and then, a new surrogate time-series is generated by exchanging temporal positions of those two segments.

Instead of the random shuffling in time domain, it can also be performed in the frequency domain as shown in Fig. 6.7c [54]. Briefly, the procedure includes fast Fourier transform (FFT), shuffling the phase of the signal in the frequency domain,

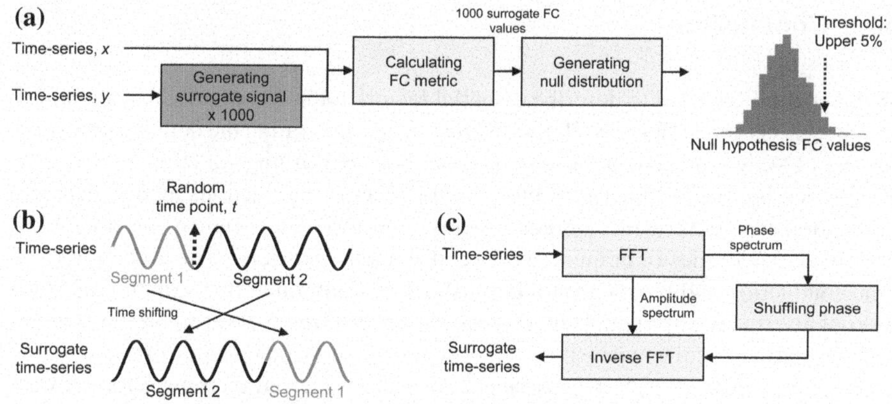

Fig. 6.7 **a** Determination of the significance of FC values using a null distribution, generated by surrogate data. **b** Generation of a surrogate time-series by time shuffling. **c** Generation of a surrogate time-series by phase shuffling (FFT: fast Fourier transform)

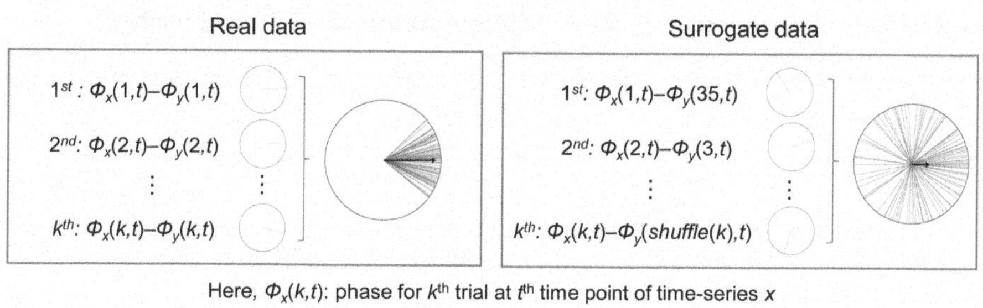

Fig. 6.8 Generation of a surrogate data by shuffling trials for an event-related data. Each grey thin solid line on the circles represents the phase difference between two time-series at each temporal point for a single trial. The black thick solid lines on the circles represent the vector sum of the phase differences over trials, and their lengths mean the phase synchronization strength

and then, the inverse FFT. The amplitude spectrum is preserved, but any nonlinear structure is destroyed after this procedure [54].

In the event-related data with a plenty of trials, shuffling the order of trials of the second time-series provides an alternative method to surrogate data [30]. Figure 6.8 shows an example for the phase-based FC metrics. For experimental recordings for which phase synchrony are expected, the phase differences between two time-series would be narrowly distributed. Contrarily, the phase differences from surrogate data would be widely distributed randomly, and thus, may provide a null distribution of FC values. This method does not require a prior hypothesis on the time-series such as linearity and stationarity, however, when the trial-to-trial variability of phase relationships between two time-series is relatively low, it can be so conservative that many FC values may be incorrectly rejected, resulting in high false negative rate.

6.7 Conclusions

In this review, we tried to describe essential factors for successful FCA using multi-channel EEG (and MEG as well) time-series. After illustrating the detailed procedure for the FCA, we presented various methods for quantifying functional connectivity. Especially, the FC measures based on oscillatory interactions among neuronal population was described comprehensively, due to its importance for elucidating coordinated activities of brain networks and synaptic plasticity. The problem of volume conduction and the means to diminish its confounding effects on the FCA was thoroughly reviewed. Spatial filtering of the time-series measured on the scalp or transformation to current densities on cortical surface, which are performed as a pre-processing for the FCA, were described. Also, we reviewed ongoing efforts toward developing FC measures which are inherently robust to the volume conduction problem. Finally, we illustrated the procedures for determining significance of the FC among specific pair of regions, which exploit surrogate data generation or the characteristics of event-related data. We hope that this review would provide guidelines for the better application of the FCA and the development of novel methods.

References

1. A.R. Anwar, M. Muthalib, S. Perrey et al., Effective connectivity of cortical sensorimotor networks during finger movement tasks: a simultaneous fNIRS, fMRI, EEG study. Brain Topogr. **29**, 645–660 (2016)
2. F. Babiloni, F. Cincotti, C. Babiloni et al., Estimation of the cortical functional connectivity with the multimodal integration of high-resolution EEG and fMRI data by directed transfer function. Neuroimage **24**, 118–131 (2004)
3. L.A. Baccalá, K. Sameshima, Partial directed coherence: a new concept in neural structure determination. Biol. Cybern. (2001). https://doi.org/10.1007/PL00007990
4. S. Baillet, J.C. Mosher, R.M. Leahy, Electromagnetic brain mapping. IEEE Signal Process. Mag. **18**, 14–30 (2001). https://doi.org/10.1109/79.962275
5. E. Barzegaran, M.G. Knyazeva, Functional connectivity analysis in EEG source space: the choice of method. PLoS ONE **12**, e0181105 (2017). https://doi.org/10.1371/journal.pone.0181105
6. A.M. Bastos, J.-M. Schoffelen, A tutorial review of functional connectivity analysis methods and their interpretational pitfalls. Front. Syst. Neurosci. **9**, 175 (2016). https://doi.org/10.3389/fnsys.2015.00175
7. R.L. Buckner, F.M. Krienen, B.T.T. Yeo, Opportunities and limitations of intrinsic functional connectivity MRI. Nat. Neurosci. **16**, 832–837 (2013). https://doi.org/10.1038/nn.3423
8. E. Bullmore, O. Sporns, Complex brain networks: graph theoretical analysis of structural and functional systems. Nat. Rev. Neurosci. **10**, 186–198 (2009). https://doi.org/10.1038/nrn2575
9. G. Buzsáki, A. Draguhn, Neuronal oscillations in cortical networks. Science **304**, 1926–1929 (2004)
10. R.T. Canolty, R.T. Knight, The functional role of cross-frequency coupling. Trends Cogn. Sci. **14**, 506–515 (2010). https://doi.org/10.1016/j.tics.2010.09.001
11. L. Canuet, R. Ishii, R.D. Pascual-Marqui et al., Resting-state EEG source localization and functional connectivity in schizophrenia-like psychosis of epilepsy. PLoS ONE **6**. https://doi.org/10.1371/journal.pone.0027863

12. J.W. Choi, K.S. Cha, J.D. Choi et al., Difficulty-related changes in inter-regional neural synchrony are dissociated between target and non-target processing. Brain Res. **1603**, 114–123 (2015). https://doi.org/10.1016/j.brainres.2015.01.031
13. J.W. Choi, K.M. Jang, K.Y. Jung et al., Reduced theta-band power and phase synchrony during explicit verbal memory tasks in female, non-clinical individuals with schizotypal traits. PLoS ONE **11**, 1–18 (2016). https://doi.org/10.1371/journal.pone.0148272
14. J.W. Choi, D. Ko, G.T. Lee et al., Reduced neural synchrony in patients with restless legs syndrome during a visual oddball task. PLoS ONE **7**, 1–9 (2012). https://doi.org/10.1371/journal.pone.0042312
15. M.X. Cohen, *Analyzing Neural Time Series Data: Theory and Practice* (The MIT Press, Cambridge, Massachusetts, 2014)
16. P.W. Diaconia, D. Freedman, Consistency of Bayes estimates for nonparametric regression: normal theory. Bernoulli **4**, 411–444 (1998)
17. K.J. Friston, Functional and effective connectivity: a review. Brain Connect **1**, 13–36 (2011). https://doi.org/10.1089/brain.2011.0008
18. G. Gómez-Herrero, M. Atienza, K. Egiazarian, J.L. Cantero, Measuring directional coupling between EEG sources. Neuroimage **43**, 497–508 (2008). https://doi.org/10.1016/j.neuroimage.2008.07.032
19. C.W.J. Granger, Investigating causal relations by econometric models and cross-spectral methods. Econometrica **37**, 424–438 (1969)
20. M. Hassan, O. Dufor, I. Merlet et al., EEG source connectivity analysis: from dense array recordings to brain networks. PLoS ONE **9**, (2014). https://doi.org/10.5281/zenodo.10498
21. S. Haufe, *Towards EEG Source Connectivity Analysis* (Berlin Institute of Technology, Berlin, Germany, 2012)
22. S. Haufe, V.V. Nikulin, K.-R. Müller, G. Nolte, A critical assessment of connectivity measures for EEG data: a simulation study. Neuroimage **64**, 120–133 (2013). https://doi.org/10.1016/j.neuroimage.2012.09.036
23. C.S. Herrmann, M.H.J. Munk, A.K. Engel, Cognitive functions of gamma-band activity: memory match and utilization. Trends Cogn. Sci. **8**, 347–355 (2004). https://doi.org/10.1016/j.tics.2004.06.006
24. A.-S. Hincapié, J. Kujala, J. Mattout et al., The impact of MEG source reconstruction method on source-space connectivity estimation: a comparison between minimum-norm solution and beamforming. Neuroimage **156**, 29–42 (2017). https://doi.org/10.1016/j.neuroimage.2017.04.038
25. J.F. Hipp, D.J. Hawellek, M. Corbetta et al., Large-scale cortical correlation structure of spontaneous oscillatory activity. Nat. Neurosci. (2012). https://doi.org/10.1038/nn.3101
26. R.M. Hutchison, T. Womelsdorf, E.A. Allen et al., Dynamic functional connectivity: promise, issues, and interpretations. Neuroimage **80**, 360–378 (2013). https://doi.org/10.1016/j.neuroimage.2013.05.079
27. M.J. Kaminski, K.J. Blinowska, A new method of the description of the information flow in the brain structures. Biol. Cybern. **65**, 203–210 (1991)
28. J. Kayser, C.E. Tenke, Principal components analysis of Laplacian waveforms as a generic method for identifying ERP generator patterns: II. Adequacy of low-density estimates. Clin. Neurophysiol. **117**, 369–380 (2006). https://doi.org/10.1016/j.clinph.2005.08.033
29. S. Khan, A. Gramfort, N.R. Shetty et al., Local and long-range functional connectivity is reduced in concert in autism spectrum disorders. Proc. Natl. Acad. Sci. U S A. **110**, 3107–3112 (2013). https://doi.org/10.1073/pnas.1214533110
30. J.-P. Lachaux, E. Rodriguez, J. Martinerie, F.J. Varela, Measuring phase synchrony in brain signals. Hum. Brain Mapp. **8**, 194–208 (1999)
31. Y.-Y. Lee, S. Hsieh, Classifying different emotional states by means of EEG-based functional connectivity patterns. PLoS ONE **9**, (2014). https://doi.org/10.1371/journal.pone.0095415
32. L. Marzetti, C. Del Gratta, G. Nolte, Understanding brain connectivity from EEG data by identifying systems composed of interacting sources. Neuroimage **42**, 87–98 (2008). https://doi.org/10.1016/j.neuroimage.2008.04.250

33. C.M. Michel, M.M. Murray, G. Lantz et al., EEG source imaging. Clin. Neurophysiol. **115**, 2195–2222 (2004). https://doi.org/10.1016/j.clinph.2004.06.001
34. T. Mima, T. Matsuoka, M. Hallett, Functional coupling of human right and left cortical motor areas demonstrated with partial coherence analysis. Neurosci. Lett. **287**, 93–96 (2000)
35. W. Mumtaz, S. Saad, A. Ali et al., A machine learning framework involving EEG-based functional connectivity to diagnose major depressive disorder (MDD). Med. Biol. Eng. Comput. (2017). https://doi.org/10.1007/s11517-017-1685-z
36. G. Nolte, O. Bai, L. Wheaton et al., Identifying true brain interaction from EEG data using the imaginary part of coherency. Clin. Neurophysiol. **115**, 2292–2307 (2004). https://doi.org/10.1016/j.clinph.2004.04.029
37. G. Nolte, A. Ziehe, V.V. Nikulin et al., Robustly Estimating the Flow Direction of Information in Complex Physical Systems. (2007). https://doi.org/10.1103/physrevlett.100.234101
38. P.L. Nunez, R. Srinivasan, A.F. Westdorp et al., EEG coherency I: statistics, reference electrode, volume conduction, Laplacians, cortical imaging, and interpretation at multiple scales. Electroencephalogr. Clin. Neurophysiol. **103**, 499–515 (1997)
39. S. Palva, J.M. Palva, The role of local and large-scale neuronal synchronization in human cognition, in *Multimodal Oscillation-Based Connectivity Theory*, ed. by S. Palva (Springer International Publishing Switzerland, 2016), pp. 51–67
40. R.D. Pascual-Marqui, Review of methods for solving the EEG inverse problem. Int. J. Bioelectromagn. Print Issue ISSN **1**:75–86 (1999)
41. R.D. Pascual-Marqui, C.M. Michel, D. Lehmann, Low resolution electromagnetic tomography: a new method for localizing electrical activity in the brain. Int. J. Psychophysiol. **18**, 49–65 (1994)
42. E. Pereda, R. Quian Quiroga, J. Bhattacharya, Nonlinear multivariate analysis of neurophysiological signals. Prog. Neurobiol. **77**, 1–37 (2005). https://doi.org/10.1016/j.pneurobio.2005.10.003
43. F. Perrin, J. Pernier, O. Bertrand, J.F. Echallier, Spherical splines for scalp potential and current density mapping. Electroencephalogr. Clin. Neurophysiol. **72**, 184–187 (1989). https://doi.org/10.1016/0013-4694(89)90180-6
44. V. Sakkalis, Review of advanced techniques for the estimation of brain connectivity measured with EEG/MEG. Comput. Biol. Med. **41**, 1110–1117 (2011). https://doi.org/10.1016/j.compbiomed.2011.06.020
45. P. Sauseng, W. Klimesch, W.R. Gruber, N. Birbaumer, Cross-frequency phase synchronization: a brain mechanism of memory matching and attention. Neuroimage **40**, 308–317 (2008). https://doi.org/10.1016/j.neuroimage.2007.11.032
46. J.M. Schoffelen, J. Gross, Source connectivity analysis with MEG and EEG. Hum. Brain Mapp. **30**, 1857–1865 (2009). https://doi.org/10.1002/hbm.20745
47. A.K. Seth, A.B. Barrett, L. Barnett, Toolbox granger causality analysis in neuroscience and neuroimaging. J. Neurosci. **35**, 3293–3297 (2015). https://doi.org/10.1523/JNEUROSCI.4399-14.2015
48. C.E. Shannon, A mathematical theory of communication. Bell Syst. Tech. J. **27**, 379–423 (1948). https://doi.org/10.1002/j.1538-7305.1948.tb01338.x
49. M. Siegel, T.H. Donner, A.K. Engel, Spectral fingerprints of large-scale neuronal interactions. Nat. Rev. Neurosci. **13**, 121–134 (2012). https://doi.org/10.1038/nrn3137
50. W. Singer, Synchronization of cortical activity and its putative role in information processing and learning. Annu. Rev. Physiol. **55**, 349–374 (1993)
51. C.J. Stam, W. De Haan, A. Daffertshofer et al., Graph theoretical analysis of magnetoencephalographic functional connectivity in Alzheimer's disease. Brain **132**, 213–224 (2009). https://doi.org/10.1093/brain/awn262
52. C.J. Stam, G. Nolte, A. Daffertshofer, Phase lag index: assessment of functional connectivity from multi channel EEG and MEG with diminished bias from common sources. Hum. Brain Mapp. **28**, 1178–1193 (2007). https://doi.org/10.1002/hbm.20346
53. C.J. Stam, J.C. Reijneveld, Nonlinear biomedical physics graph theoretical analysis of complex networks in the brain. Nonlinear Biomed. Phys. **1**, 1–19 (2007). https://doi.org/10.1186/1753-4631-1-3

54. J. Theilier, S. Eubank, A. Longtin et al., Testing for nonlinearity in time series: the method of surrogate data. Phys. D **58**, (1992)
55. M.P. Van Den Heuvel, H.E. Hulshoff Pol, Exploring the brain network: a review on resting-state fMRI functional connectivity. Eur. Neuropsychopharmacol. **20**, 519–534 (2010). https://doi.org/10.1016/j.euroneuro.2010.03.008
56. F. Varela, J.-P. Lachaux, E. Rodriguez, J. Martinerie, The brainweb: phase synchronization and large-scale integration. Nat. Rev. Neurosci. **2**, 229–239 (2001)
57. F. Vecchio, F. Miraglia, C. Marra et al., Human brain networks in cognitive decline: a graph theoretical analysis of cortical connectivity from EEG data. J. Alzheimer's Dis. **41**, 113–127 (2014). https://doi.org/10.3233/JAD-132087
58. M. Vinck, R. Oostenveld, M. Van Wingerden et al., An improved index of phase-synchronization for electrophysiological data in the presence of volume-conduction, noise and sample-size bias. Neuroimage **55**, 1548–1565 (2011). https://doi.org/10.1016/j.neuroimage.2011.01.055
59. B. Voytek, M. D'esposito, N. Crone, R.T. Knight, A method for event-related phase/amplitude coupling. Neuroimage **64**, 416–424 (2013). https://doi.org/10.1016/j.neuroimage.2012.09.023
60. L.M. Ward, Synchronous neural oscillations and cognitive processes. Trends Cogn. Sci. **7**, 553–559 (2003). https://doi.org/10.1016/j.tics.2003.10.012

Part III
Applications

Chapter 7
Computational EEG Analysis for the Diagnosis of Psychiatric Illnesses

Seung-Hwan Lee and Yeonsoo Park

Abstract Electroencephalography (EEG) holds promise as a tool to diagnose psychiatric disorders. While it has some major advantages such as high temporal resolution, relative affordability, and easy accessibility, even its shortcomings are being addressed through the advancement in its analysis. As a result, numerous researches have been examining EEG components as potential biomarkers of various psychiatric diseases. In this chapter, we discuss several promising EEG markers, ranging from resting state EEG to stimuli induced ERP components, from electrodes level to source level, and from band power to functional connectivity networks. In addition, we present the findings of previous studies with an emphasis on how each EEG component vary depending of the specific psychiatric illnesses. The psychiatric disorders discussed in this chapter are (1) schizophrenia, (2) bipolar disorder, (3) major depressive disorder, (4) anxiety related disorders (e.g., post-traumatic stress disorder and obsessive compulsive disorder) and (5) disorders related to cognitive impairments (e.g., dementia and minimal cognitive impairment. Lastly, we introduce how the limitations of EEG, which mostly occur as a byproduct of sensor-level analysis, can be addressed through source-level analysis.

7.1 Introduction

Traditionally, psychiatric disorders have been diagnosed primarily through face-to-face interviews, and biological measures that could effectively capture clinical features of mental illnesses were relatively scant. To overcome such unmet needs, many researchers placed meticulous effort into developing biomarkers that could reliably detect the clinical characteristics of specific psychopathologies or psychiatric illnesses. Among the various biomarkers, electroencephalography (EEG) is a

S.-H. Lee (✉)
Department of Psychiatry, Inje University, Ilsan-Paik Hospital, Goyang, Republic of Korea
e-mail: lshpss@paik.ac.kr

S.-H. Lee · Y. Park
Clinical Emotion and Cognition Research Laboratory, Inje University, Goyang, South Korea

C.-H. Im (ed.), *Computational EEG Analysis*, Biological and Medical Physics, Biomedical Engineering, https://doi.org/10.1007/978-981-13-0908-3_7

promising approach with high potential in measuring the pathologies of mental illnesses. It possesses several crucial merits, such as high temporal resolution, relatively low economic burden, and easy accessibility compared to other neuroimaging methods [e.g., magnetic resonance imaging (MRI), positron-emission tomography (PET), and magnetoencephalography (MEG)].

During the past few decades, the EEG analysis methods have rapidly evolved in line with the advancement of technology in fields such as engineering and physics. While EEG analysis at the electrode level (surface of scalp) had critical limitations as neuronal signals of the brain would blur or fade as they reached the scalp, the development of source level analysis methods has brought an evolution by allowing signals to represent the characteristics of the cortical level. This evolution has enabled researchers to conduct more sophisticated studies that inspect region specific biomarkers, because the source level analysis could use the information of brain anatomical region. Currently, we are at the midpoint of developing a variety of future biomarkers that accurately portray pathological features. During this stage, the development of EEG biomarkers is expected to play a pivotal role.

In this chapter, we will first review the candidate EEG biomarkers that have been studied in relations to various psychiatric disorders. General characteristics of these EEG biomarkers will be introduced. In the following section, we will look into specific disorders and discuss how the previously mentioned EEG biomarkers manifest to make them reliable markers of specific mental illnesses.

7.2 Promising EEG Biomarker Candidates

7.2.1 P300

P300 is the most widely studied event-related potential (ERP) component reflecting brain functions such as attention, working memory, and cognitive decline. It appears most ostensibly at the centro-parietal scalp areas and is usually elicited within 250–500 ms after the stimulus presentation. There are unique distinctions between two regionally dependent P3 components, specifically the frontally maximal P3a component and the parietally maximal P3b component. The P3a component is triggered by infrequent distinct tones presented within a series of frequent tones during a resting state. The P3b component is a task-relevant potential elicited during target stimulus processing. The P3a component has been associated with attention mechanisms and novel stimulus processing, while P3b has been more related to stimulus evaluation and decision making [111]. When experts of neurophysiology mention the P3 component, they usually refer to the P3b [84, 111].

Because of its reflections of cognitive functioning, P3 components have been widely studied in respect to psychiatric illnesses that are characterized by cognitive impairments. Although most studies have been confined to the electrode level, changes of amplitude and latency of P3 have been confirmed in various neuropsychiatric disorders.

7.2.2 Functional Connectivity Network

Functional network analysis can be conducted on a wide range of EEG data. One material that can be used universally to examine functional connectivity is the resting state EEG. Resting state EEG holds many advantages because it is easy to obtain, relatively stable, not difficult to handle, and independent from task characteristics. Yet, several inevitable qualities require researchers to be cautious when dealing with resting state EEG. For example, resting state EEG that is measured during an eyes-closed condition can contain unwanted artifacts, which result from drowsiness. This leads to questions regarding the length and the number of epochs required in order to obtain stable statistical power and how expert consensus on the standardization for data qualification ought to be driven. While these points can pose some threat as to the reliability of the measure, they are not insoluble and can be adequately addressed [33]. Resting state EEG clearly has more advantages than weaknesses. Hence, EEG functional connectivity of the resting state will be mainly discussed throughout this text.

An increasing number of researchers have assumed that alterations in the cortical connectivity network might provide insights to the underlying neural mechanisms of mental illnesses. Many of these studies adopted the graph theory to quantify global and local changes in the cortical functional connectivity network [12, 115, 121, 125]. In particular, the small-world network has been regarded as one of the most suitable models to elucidate information transfer in the human brain [10]. The small-world network is the middle ground between random network and regular network, and is characterized by a higher clustering coefficient than random networks, and a shorter path length than regular networks. The clustering coefficient and the path length each reflect the amount of segregation of highly inter-connected units and the amount of integration of the whole network [140]. Therefore, the small-world characteristics of the brain allow for more efficient information transfer among distant brain regions.

On the other hand, a weighted network is a network where the ties among nodes have weights assigned to them. Because of this nature, weighted networks are more difficult to analyze than unweighted binary networks in which ties are simply present or absent. Despite the difficulty of its analysis, a number of network measures have been proposed for weighted networks including the following:

- Strength, which refers to the strength of the connection in the network, is estimated by the sum of weights of links connected to the brain regions.
- Clustering coefficient (CC), which refers to the degree in which a node is clustered with neighbor nodes, is calculated for the whole network.
- Path length (PL), which refers to the summation of lengths between two nodes in the whole network, indicates overall connectedness of the whole network.

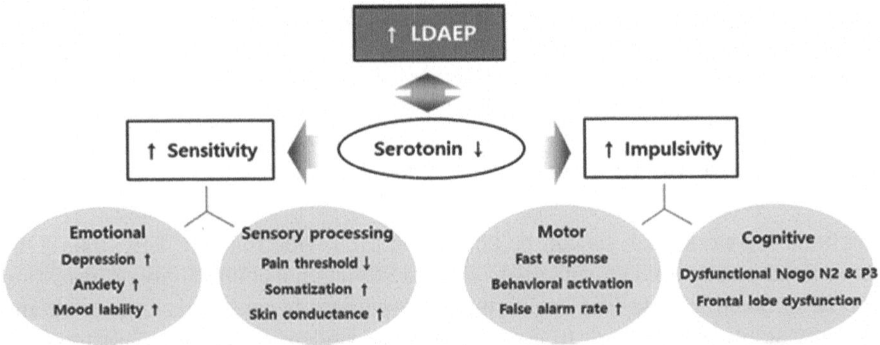

Fig. 7.1 Loudness dependence of auditory evoked potential (LDAEP) could reflect sensitivity and impulsivity because both reflect serotonin related regulation. Theoretically, LDAEP could be correlated with emotional and sensory processing sensitivity. Moreover, LDAEP might be associated with impulsivity, which influences motor and cognitive systems. Cited from the article of Kim et al. [62]

- Efficiency refers to the efficiency of information processing in the brain.

7.2.3 Loudness Dependence of Auditory Evoked Potential

The loudness dependence of auditory evoked potential (LDAEP) has been suggested as a valid marker of central serotonergic activity. Empirical studies have found significant inverse correlations between LDAEP and central serotonergic activity, indicating that high LDAEP reflects low levels of serotonergic neurotransmission [53]. The LDAEP is measured through calculating the amplitude change of the evoked N1/P2 components in response to auditory stimuli that vary in intensity [38], O'neill et al. [97]. Greater amplitude changes reflect sensitivity to auditory stimuli, which then again corresponds to the strength of LDAEP. Based on their associations with central serotonin activity, the relationship between LDAEP and sensitivity (i.e., emotional and sensory) and impulsivity has been studied [62] (see Fig. 7.1).

Both sensitivity and impulsivity not only are central characteristics of various psychiatric disorders but they also heavily influence their progression. Impulsivity is defined as the inability to inhibit inappropriate behaviors [6, 126]. Previous literature has reported that the serotonin system plays a critical role in behavioral inhibition and impulsivity [41, 85, 126], and dysregulations of the serotonin system in sub-regions of the prefrontal cortex are associated with maladaptive behaviors. Considering the associations between impulsivity and central serotonin system, and serotonin activity and LDAEP, a significant relationship between LDAEP and impulsivity can be assumed.

Although, the relationship between sensory sensitivity and impulsivity is not conclusive and might differ in various clinical populations, a relationship between the two constructs can be assumed when considering the serotonin system as a mediator. Therefore, more studies examining LDAEP as a candidate of clinical markers for sensitivity and impulsivity demand greater clinical attention.

7.2.4 The 40-Hz Auditory Steady-State Response

The 40-Hz auditory steady-state response (ASSR) measures oscillatory responses following auditory stimulation at gamma-band frequency [130]. Gamma-band oscillations play an important role in establishing temporal precisions in local cortical networks [133], and are candidate mechanisms of perceptual integration, attentional selection, and working memory [116, 133]. Because of its importance in a variety of cognitive functions, abnormal gamma oscillations have been actively studied in schizophrenia and bipolar disorder, which include a wide set of cognitive and perceptual anomalies as their primary symptoms [98, 100, 112].

Abnormal gamma oscillations are a critical element in current disease models of schizophrenia, including *N*-methyl-D-aspartate (NMDA) receptor hypofunction and altered excitation and inhibition balance models. Indeed, the generation of gamma oscillations has been shown to depend on the integrity of neural circuits involving fast-spiking parvalbumin-expressing cells, and the hypofunction of the NMDA receptor on these interneurons has been suspected to underlie psychosis [29, 55, 77]. Importantly, pharmacological and genetic animal studies imply that mechanisms of gamma oscillation generation are relatively well-articulated and may converge across species, rendering gamma oscillation, and consequently the 40-Hz ASSR a potentially promising target of translational research [88].

7.2.5 Mismatch Negativity

Mismatch negativity (MMN) is an automatically generated ERP component when a sequence of relatively uniform stimuli is interrupted by the infrequent presentation of deviant stimuli. Initially discovered in an auditory experimental procedure, MMN is thought to reflect an automated detection of perceptual change or a sensory prerequisite of cognition and has been closely examined in relation to attentional processes [94]. In particular, the MMN latency defines the response time change in the sensory stimulus [131].

Although MMN can occur in any sensory system, it has been most frequently studied in the visual and the auditory systems. In the case of auditory stimuli, MMN occurs after an infrequent change in a repetitive sequence of sounds. The deviant sound can differ from the standards in one or more perceptual features such as pitch, duration, or loudness. In addition, MMN is usually evoked either by a change in frequency, intensity, duration, and by apparent spatial locus of origin. Deficiency in MMN appears to be an index of cognitive decline, irrespective of the specific symptomatology and etiologies of the involved disorders [95]. Source analysis of MMN in EEG and MEG signals localizes MMN to the auditory cortex, adjacent areas of the superior temporal lobe, middle and inferior frontal gyrus, and anterior cingulate cortex.

Furthermore, MMN has several clinical implications. For example, it correlates with disease severity, cognitive dysfunction, impaired social functioning, and brain alterations such as reductions in gray matter. Reductions in MMN may imply deficits in NMDA receptors, which function abnormally in patients with schizophrenia. It has been reported that individuals with chronic schizophrenia exhibit reduced MMN amplitudes, 0.94 standard deviation (SD) smaller for pitch deviants and 1.23 SDs smaller for duration deviants, compared with healthy controls. Moreover, MMN deficits also appear to be more severe in schizophrenia compared to individuals with bipolar disorder or Alzheimer's disease.

7.2.6 Alpha Asymmetry

Alpha asymmetry has been extensively studied in both clinical and nonclinical samples. Among various psychiatric disorders, it has been studied mostly in depression and anxiety disorders. Alpha asymmetry is characterized by an asymmetrical alpha-band activity (8–12 Hz) in the left and right hemispheres [36], and known to be a reflection of cortical activity [18]. Thus, increased alpha indicates decreased cortical activity and decreased alpha indicates increased cortical activity. Studies have identified alpha asymmetry mostly in the frontal and the parietal regions of the brain.

Perhaps, the most referred explanation of the relationship between alpha asymmetry and psychiatric disorders is the behavioral activation/inhibition systems (BAS/BIS). Initially proposed by Gray to understand the psychophysiology of personality configurations [32], the BAS/BIS have been applied to studies examining different psychopathologies. For example, empirical research has suggested that depression is characterized by hypofunctioning of the BAS, which signifies that depressive individuals demonstrate an indifference towards positive stimuli [56]. Davidson argued that both the BAS and BIS are lateralized hemispherically, whereby the left hemisphere regulates the BAS and the right hemisphere regulates the BIS [19]. Results supporting the relationship between alpha asymmetry and BAS have been more constant that those between alpha asymmetry and BIS [16, 36, 129]. There are still inconstant reports about alpha symmetry and major depressive disorder. Comorbid anxiety disorder, subtype of depressive disorder considering individual variance of characteristics in terms of BIA/BAS, and the presence of suicidal ideation should possibly be considered to obtain more coherent findings in future. Nonetheless, alpha asymmetry seems to have promising value as a marker of psychiatric disorders.

7.3 Psychiatric Disorders and EEG Biomarkers

7.3.1 *Schizophrenia*

7.3.1.1 Auditory Steady-State Response (ASSR)

As was mentioned in the previous section, abnormal gamma oscillation is an important biomarker of schizophrenia; it is thought to be related to NMDA receptor hypofunction and altered excitation and inhibition balance models. Most of human studies on 40-Hz ASSR reported reductions of evoked power and phase measure in schizophrenia [130]. However, these results are not entirely consistent. Hong and colleagues rather found enhanced evoked powers in patients with schizophrenia who were taking new generation antipsychotics [43]. Hamm et al. also reported higher phase measure in patients with schizophrenia compared with healthy controls, where they employed novel stimulus parameters including broadband auditory noise and wide inter-stimulus interval (ISI) [35]. Notably, a new line of research investigating the level of spontaneous (non-phase locked) gamma-band activity, found enhanced gamma activity during both ASSR stimulus presentation and task baseline, but not in a resting-state [42]. This finding could be significant because it resonates with animal findings where NMDA receptor antagonists and genetic reduction of NMDA receptor function induced increases in spontaneous gamma power [22, 66].

A reason underlying the inconsistency might be the obscure definition of ASSR components as researchers use different ASSR components from classifications that have not reached consensus. Recently, Mathalon and Sohal systematically classified the ASSR components into stimulus-evoked oscillations (evoked power), stimulus-induced oscillations (total power), and stimulus independent oscillations (baseline or resting-state power) [86]. This classification holds importance because the signals reflect different aspects of information processing. For example, the evoked power reflects bottom-up sensory encoding, while spontaneous gamma (the stimulus-induced and stimulus independent oscillations) are related to emerging dynamic processes in cortical networks [46]. Although the argument to distinguish ASSR signals has been already established, only a few studies have explicitly used it to date. In order to reduce the inconsistency in the results, future research ought to take this into consideration.

Considering that cortical pyramidal cell and interneurons are major sources of gamma band oscillations [133], another important question is how the 40-Hz ASSR would be associated with well-known neuropathological features of schizophrenia such as gray matter loss and symptoms [34]. A major brain area thought to be involved in generating the 40-Hz ASSR is the primary auditory cortex located in the superior temporal gyrus (STG) [40]. Recently, Kim and colleagues reported that the ASSR gamma power was significantly increased in patients with schizophrenia compared with healthy controls [63] (see Fig. 7.2). The discrepancy about ASSR gamma power might have occurred because the inter-train interval of ASSR stimuli train was different among studies. The study of Kim and colleagues used longer

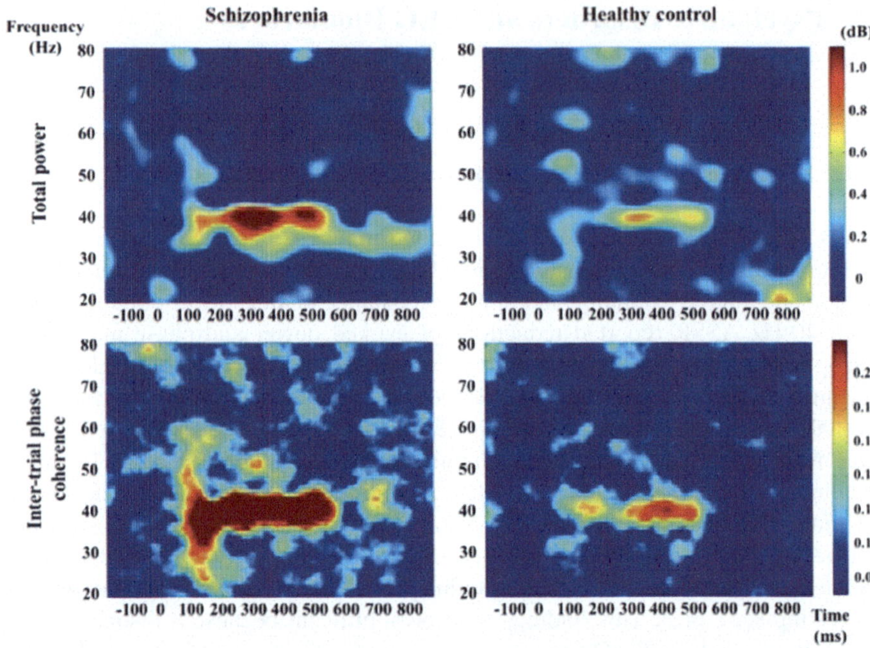

Fig. 7.2 The grand average of time-frequency maps of total power and inter-trial phase coherence at Cz in the 40 Hz frequency for patients with schizophrenia and healthy controls. In contrast to most earlier results, total power and inter-trial phase coherence were significantly increased in patients with schizophrenia compared with healthy controls. Inter-train interval (3050–3500 ms) was used in this study, which was longer than that used in most previous ASSR studies

inter-train interval (3000 ms), while others used rather short ISI (less than 1000 ms). Longer ISI might contribute to generating the gamma band response by modulating the refractory time of pyramidal cells and interneurons in patients with schizophrenia.

7.3.1.2 Functional Connectivity

In previous EEG studies, patients with schizophrenia consistently showed disrupted small-world networks characterized by decreased clustering coefficients and prolonged path lengths in the resting state [47, 91, 114] and during working memory tasks [119]. However, most EEG network analysis, including the examples mentioned above, were confined to connectivity analysis at the electrodes (sensor) level. Therefore, these studies failed to report the specific cortical regions that contribute to disruptions of the small-world cortical functional network. EEG topographies cannot be directly matched with the underlying cortical regions, since electrodes (sensors) may contain information from multiple brain sources, some of which might overlap. In addition, topographic maps are sometimes smeared out due to inhomogeneous

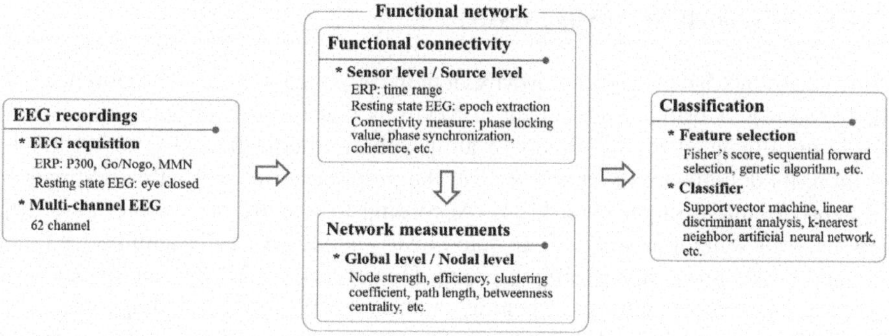

Fig. 7.3 Schematic sequence of psychiatric disease classification using EEG. The EEG signals can be transformed into functional connectivity indices, and the functional connectivity index can be subsequently recalculated to network measures. After feature selection, classifiers could be applied for differential diagnosis of psychiatric illness

conductivity distributions in the human head. This so-called *volume conduction effect* can cause spurious connectivity between scalp EEG channels [37], eventually leading to failure in identifying the region-specific changes in functional connectivity networks. This shortcoming has been addressed by Fallani et al., who performed the first network analysis (node degree and network density) of EEG source-level functional connectivity in patients with schizophrenia during the 2-back working memory task [24].

A noteworthy limitation of previous studies on functional connectivity in schizophrenia patients is that the majority of studies applied binary (unweighted) functional networks to estimate small-worldness. This method utilizes arbitrary threshold values to convert the original functional connectivity network into a binary form. During this process, information regarding the strength of interactions potentially useful in identifying small-world characteristics in patients with schizophrenia, can be lost. Therefore, weighted functional networks is necessary to obtain more realistic functional networks in schizophrenia.

Previously, our group published a small-world cortical functional connectivity network during an auditory oddball paradigm task in patients with schizophrenia [122]. The results suggested that the small-world functional network is disrupted in patients with schizophrenia. Moreover, the negative and cognitive symptom components of positive and negative symptom scales were negatively correlated with the clustering coefficient and positively correlated with path length. With these information about brain anatomy-based knowledge, further research should be conducted for distinguishing patients with schizophrenia from those with other psychotic mental illnesses. To accomplish this, the machine learning and deep learning technology would be used as pivotal tools of classifier with the feature selection from various clinical data (see Fig. 7.3).

7.3.1.3 Mismatch Negativity (MMN)

Multiple studies have identified MMN deficits in patients with schizophrenia [79, 80, 117, 141]. A primary advantage of MMN is that it has been known to be relatively uninfluenced by the effects of antipsychotic medication [14, 135, 136]. In addition, its deficits are thought to reflect the progression of a disease and premorbid neurocognitive impairment [134]. According to a study by Şevik et al. [120], patients with schizophrenia demonstrated nearly identical MMN amplitudes to an age- and education-matched sibling group. However, their MMN amplitudes were significantly lower than those of healthy controls [120]. Similar findings were also reported by Lee et al. [72, 74], who showed that patients with schizophrenia exhibited comparable MMN amplitudes compared to that of first degree relatives, but significantly lower than education-matched healthy controls. In addition, the MMN amplitude of frontal electrodes and functional outcomes measurements showed the most powerful correlations compared to other psychological measurements in patients with schizophrenia.

The major pathology of MMN deficit appears to originate from dysfunctions of the NMDA receptor system [48, 137]. NMDA-receptor-mediated glutamatergic dysfunction may well explain the pathology of both schizophrenia and other neuropsychiatric diseases, which explicitly reflect MMN deficits [137].

Several studies have reported correlations between MMN and global social functioning in patients with chronic schizophrenia [57, 59, 72, 74, 79, 80], and one study found a stable association over a 1-year period [79, 80]. A significant association between MMN and Global Assessment of Functioning (GAF) scores have also been noted in healthy participants [81], indicating that deficits in MMN can impact social functioning not only in patients but in community samples as well.

Studies measuring MMN reduction in patients with first-episode schizophrenia-spectrum showed a negligible effect size of 0.04 SD for MMN to a pitch-deviant and a small to medium effect size of 0.47 SD for MMN to a duration-deviant. Effect sizes for MMN reductions in patients with chronic schizophrenia were around 1 SD compared with controls, suggesting that the MMN deficit increases with the progression of the disease. Despite the marked MMN reductions in patients with chronic schizophrenia, the deficit does not seem to be severe during the first episode. There is no consistent evidence for a marked deficit in pitch MMN in patients with first-episode schizophrenia spectrum, while MMN may show a small to medium effect size along with the progression of the disorder (Fig. 7.4).

7.3.2 Bipolar Disorder

Bipolar disorder is a major psychiatric illness characterized by recurrent manic, depressive, mixed, and euthymic phases. Emotional dysregulation and cognitive impairments, which possibly reflect structural and functional brain connectivity disturbances [11, 15], seem to be involved in the disorder [68]. Alterations in white mat-

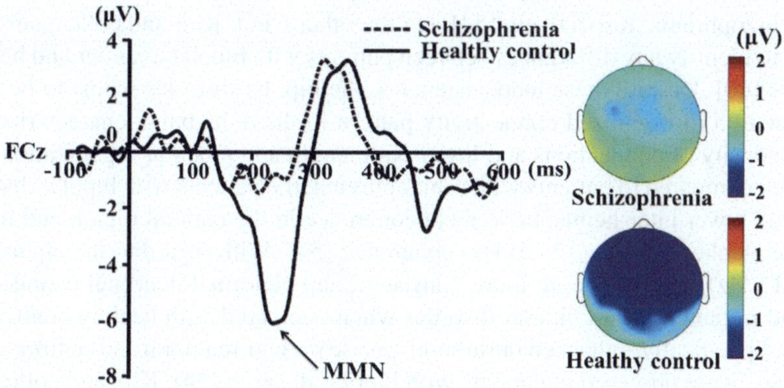

Fig. 7.4 Topographic maps of MMN and MMN waves at FCz in patients with schizophrenia and healthy controls. MMN, mismatch negativity

ter frontolimbic circuitry, inter-hemispheric connectivity, and fronto-parietotemporal connections, which have been reported in patients with bipolar disorder [11], may support the emergence of functional disturbances. However, direct evidence for neural circuitry abnormalities in bipolar disorder is still inconclusive.

Studies have mostly focused on examining intra- and inter-hemispheric EEG connectivity in patients with bipolar disorder in comparison to patients with schizophrenia and healthy controls. The comparisons have been usually conducted either in resting state or during visual or motor tasks. Representative measures of connectivity have been EEG coherence and synchronization likelihood (SL). The EEG coherence measures the extent of linear oscillatory coupling between two signals [54], and SL is sensitive to linear and nonlinear dependencies between two signals [93]. In addition, some studies extracted topological network parameters (e.g., modularity, path length, and clustering coefficient), which allow the identification of key organizational principles governing the brain networks [12]. However, most of the studies did not account for the possible effects of pharmacological therapies on electrophysiological activity of patients with bipolar disorder.

Studies examining the patterns of EEG coherence in the delta, theta, alpha, and beta frequency bands have been usually conducted during the resting state. The only results that were consistent across frequency bands were reported by Bhattacharya [9], who found a lower degree of long-range phase synchrony in patients with manic bipolar disorder than in healthy controls [9]. Other studies have rather produced inconsistent results. In Barttfeld et al. [7], patients with euthymic bipolar disorder showed higher SL values than healthy controls, especially among frontal and occipital cortices [7]. Additionally, patients with bipolar disorder exhibited altered network topology with higher nodal degree, but shorter characteristic path length and lower modularity than healthy controls. Kam and colleagues found reduced frontal interhemispheric EEG coherence and reduced intrahemispheric EEG coherence in patients with bipolar disorder patients compared with healthy controls and patients with schizophrenia, respectively [54]. On the other hand, Kim and colleagues were unable to identify any differences between patients with bipolar disorder and healthy controls [60]. Despite these inconsistencies, the bipolar disorder seems to be associated with a disorganized connectivity pattern in the delta band, characterized by higher density of connections and lower path length and modular organization.

When compared to patients with schizophrenia, the patients with bipolar disorder exhibited lower inter-hemispheric theta coherence in the parietal region and higher intra-hemispheric beta1 (12–20 Hz) coherence [54]. With regard to the alpha band (8–12/13 Hz), the results are more consistent and abnormal neuronal connectivity is found in patients with bipolar disorder when compared with healthy controls [9, 54, 60]. Specifically, enhanced intra-hemispheric parieto-temporal and centroparietal coherence were observed in patients with bipolar disorder [54]. Kim and colleagues showed that patients with bipolar disorder had lower SL values and nodal strength, especially in a fronto-central-parietal network [60]. In addition, patients with bipolar disorder had lower clustering coefficient and global efficiency, but higher values of characteristic path length than healthy controls. Similarly, Bhattacharya observed the strongest reduction in the degree of long-range phase synchrony in the alpha band [9]. Furthermore, alpha EEG coherence abnormalities seem to persist even in patients with bipolar disorder who are in an active phase (manic and/or depressive) [101].

Özerdem and colleagues used a visual oddball paradigm to investigate the event-related gamma coherence in patients with manic [104] and euthymic [103] bipolar disorder. The authors showed that patients with manic bipolar disorder exhibited reduced fronto-temporal coherence in the right hemisphere. Interestingly, this difference was also confirmed in patients with euthymic bipolar disorder. In Valesques et al. [138], gamma EEG coherence was explored in patients with depression and manic bipolar disorder and manic bipolar disorder alone compared with healthy controls during a prosaccadic paradigm [138]. Interestingly, patients with manic bipolar disorder exhibited higher gamma coherence in the bilateral frontal region, whereas patients with depression and bipolar disorder showed increased coherence in the right frontal region in comparison with healthy controls. These results suggest that abnormal information processing in bipolar disorder may rely either on increased or decreased gamma coherence, depending on the functionality of each brain region during the examined task.

In a recent review, Özerdem et al. reported that (a) the decrease of higher frequency theta (6–8 Hz) response might reflect impaired cognition in bipolar disorder, (b) breaks of spontaneous alpha oscillation might indicate a deficit of the central nerve system in bipolar disorder, and (c) beta oscillations might reflect treatment response in bipolar disorder [102] (see Fig. 7.5).

Illness state / Method	Euthymic								Manic							
	Drug-free				Lithium treated				Drug-free				Valproate treated			
	θ	α	β	γ	θ	α	β	γ	θ	α	β	γ	θ	α	β	γ
Power spectrum																
Spontaneous EEG		⬇														
Sensory evoked		↓														
Coherence																
Event related				⬇								⬇				↓
Evoked				↓								↓				↓
Event-related oscillatory response	⬇		↑				⬆			⬇	⬆			⬇	↔	

Fig. 7.5 Results from potential electrophysiological biomarkers in bipolar disorder. All depicted comparisons were made with healthy controls. Thickness of arrows indicate significance of the difference (Thick, significant; Narrow, non-significant). The direction of the arrows indicates increase, decrease, or similar results between patients with bipolar disorder and healthy controls. Cited from the article of Ozerdem et al. (Supplements to Clinical neurophysiology. 2013;62:207–21) [102]

7.3.3 *Major Depressive Disorder*

Major depressive disorder is associated with high lifetime prevalence, estimated between 13.2 and 16.5% [139]. Its socio-economic burden and, more importantly, its impact on the quality of life of the patients have triggered efforts to identify biomarkers that might help to predict better its prognosis and response to treatment.

7.3.3.1 Alpha Asymmetry

Alpha asymmetry of the frontal lobe might possibly be one of the most frequently studied biomarkers of major depressive disorder. Empirical research has constantly reported a greater relative alpha power in the left hemisphere than the right in patients with major depressive disorder [31, 39]. Conversely, cortical activation has been known to appear in an opposite pattern with alpha activity, and depressive individuals typically demonstrate a hypoactivation in the left frontal hemisphere [20, 31, 49].

Attempts to understand the significance of alpha asymmetry in patients with major depressive disorder have been conducted through the BAS and the BIS. The left frontal hemisphere has been postulated to be operated by the BAS, which oversees approach behaviors in the presence of rewards or incentives. The depressive symptoms of major depressive disorder might be attributable to a hypofunctioning left frontal hemisphere, which is reflected by asymmetrical increased alpha activity of left frontal region [19]. Empirical research has been in support of such hypothesis as increased relative alpha activity in the left hemisphere has been associated with decreased scores on self-reported questionnaires of BAS in depressive patients [21, 25].

Although theoretically supported, there are several confounding factors that ought to be regarded in further studies on alpha asymmetry. Foremost, frontal alpha asymmetry is assumed to reflect the activity of the frontal lobe and any biological conditions that could influence its functioning should be controlled in order to properly examine the effects of alpha asymmetry. Several conditions have been suggested to influence frontal lobe functioning in major depressive disorder. First the subtypes of depression (for example, melancholic *vs.* atypical depression) should be controlled due to its different pathophysiology in relation to frontal lobe function. Second, the various symptom manifestations (e.g., suicide ideation) that might influence frontal lobe function ought to be considered. Frontal lobe function could be significantly influenced by suicidal ideation. The disturbed functioning as well as reduced volumes of the orbitofrontal cortex have been reported to be associated with increased impulsivity and aggression, which are vulnerabilities of suicidal behavior [87, 92]. Third, comorbid disorders ought to be regarded as different psychiatric disorders are thought to have distinct alpha asymmetry patterns. For example, patients with anxiety disorders often reveal an opposite pattern of alpha asymmetry in that greater relative alpha power in the right hemisphere than the left [25].

7.3.3.2 Loudness Dependence of Auditory Evoked Potential (LDAEP)

The LDAEP has been related to depression, anxiety, and mood lability in clinical samples as well as healthy participants. Several studies have suggested that sensory processing sensitivity is associated with depression and anxiety. More specifically, patients with anxiety or depression reported higher pain sensitivity, skin conductance, or somatic sensations. Considering that LDAEP might reflect sensitivity (i.e., emotional and sensory) through its relations to central serotonin systems [62], LDAEP might also reflect vulnerabilities toward depression.

The LDAEP has not only been associated with sensitivity but also with impulsivity. Previous studies reported that impulsive persons demonstrated faster reaction time and more false alarm rates in a Go/No-go task. The LDAEP appears to reflect impulsivity as the changes of No-go P3 amplitude correlated with LDAEP [62]. Additionally, LDAEP has been postulated to reflect impulsivity as represented by motor and cognitive inhibition. An increasing number of researches have been supporting the relationship among LDAEP, impulsivity, and sensitivity, suggesting the possibility of LDAEP as a valid biomarker of psychopathology.

In addition, LDAEP could be a trait-like state marker of mood disorder. While LDAEP showed a similar pattern between patients and healthy controls, it was known to predict treatment response of mood and anxiety symptom that were considered serotonin disturbances in their background pathology. There is a significant correlation between a strong LDAEP—indicating low serotonergic function—and

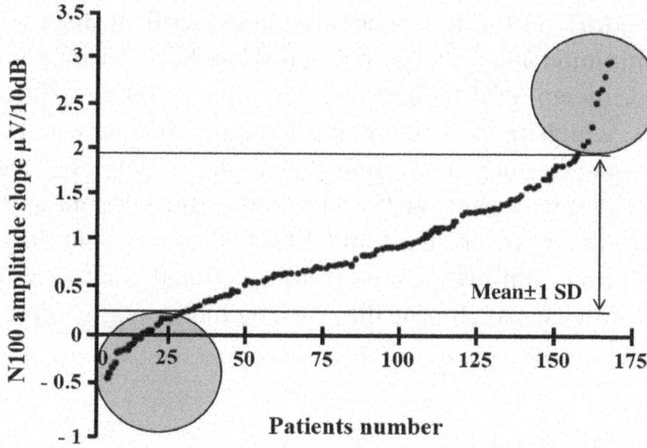

Fig. 7.6 Distribution of the N 100 amplitude slope (LDAEP) in 166 patients with major depression (38 men and 128 women) aged 47.6 ± 18.4 years. Individuals with lower LDAEP could have lower antidepressant responsiveness because they have relatively high level of serotonergic activity, while individuals with high LDAEP could have better antidepressants responsiveness, even induced hypomania in extreme cases, because they have relatively low level of serotonergic activity. Cited from the article of Seung-Hwan Lee et al. with the permission of Korean College of Neuropsychopharmacology [73]

a favorable response to SSRIs in patients with major depressive disorder and generalized anxiety disorder [69, 108]. In addition, there is some evidence that LDAEP could reflect the treatment response of mood disorder including lithium treatment [52]. For example, patients with an extreme level of LDAEP (too high or too low) could show aberrant response for antidepressants treatment or treatment responsiveness for antidepressant or lithium treatment (Fig. 7.6).

7.3.3.3 Functional Connectivity

Most EEG studies of major depressive disorder have provided information about the electrode (sensor) level's coupling between distinct cortical areas with non-linear (phase synchronization) and linear (amplitude) properties of the time series that allow analysis of relevant network activity in EEG data.

Some early studies reported differences of the EEG coherence, between patients with major depressive disorder and healthy controls [78, 96]. Since then, different measures (e.g., partial directed coherence, Granger causality, structural synchrony index, phase synchrony index) have emerged that might help to assess alterations of EEG-based connectivity in major depressive disorder. Multiple studies reported a decreased EEG coherence measures in major depressive disorder [64, 75, 106, 128]. In contrast, other studies found an increased EEG connectivity in major depressive disorder, most consistently in the alpha band [26, 50, 76]. In order to substantiate functional connectivity as a robust indicator of depression, research ought to clarify why the direction of the alterations (i.e., decrease or increase of connectivity measures) differ depending on the independent studies. Critical discussions about the impact of EEG amplitude and volume conduction on the different connectivity measures might provide meaningful suggestions. The alpha amplitude shows big variance among individuals, which may reflect unrevealed individual personal characteristics. If the researchers want to study the resting state alpha activity, the sleepiness during EEG measurement should be rigorously controlled to rule out contamination of alpha power by smearing theta power effect on EEG. If these confounding factors could be well controlled, more uniform results from functional connectivity and network analysis would be drawn in psychiatric illnesses, including major depressive disorder.

7.3.4 Anxiety Disorder

Most EEG studies have reported alterations of multiple spectral bands in anxiety disorders. Alterations of the theta (4–8 Hz) and alpha (8–13 Hz) bands are often observed [5]. However, the beta and gamma band abnormality were also recently reported in patients with post-traumatic stress disorder (PTSD) [72, 74, 121]. None of the qEEG alterations are refrained to anxiety disorders. However, they are thought to be related to symptoms of anxiety disorders and have been targets of neurological treatments such as neurofeedback training [124]. Among the various anxiety disorders, PTSD and obsessive-compulsive disorder are some that have received the most empirical attention.

7.3.4.1 Post-traumatic Stress Disorder (PTSD)

Similar to major depressive disorder, frontal alpha asymmetry is a frequently studied biomarker of PTSD [30, 58]. Relatively greater left frontal activity is regarded as being related to appetitive motivation, and lower levels of depressive and anxious symptoms in patients with PTSD [90]. However, the implication of alpha asymmetry in PTSD is not definitive; a study comparing alpha asymmetry between patients with PTSD and healthy controls reported no significant differences [30].

Connectivity measures have also been to be of interest when studying EEG biomarkers of PTSD. Patients with PTSD, when compared with controls, were found to have decreased resting-state EEG frontal connectivity, especially in beta and gamma frequency bands. In addition, nodal values (connection strength) of these two frequency bands were significantly correlated with PTSD symptom severity, and with depression and increased arousal [72, 74].

In addition, significant associations have been found between PTSD symptoms, and P200 and P300 ERP components [82]. A recent study by Shu et al. reported that alterations of ERP components (N200 and P300 amplitudes) accurately classified whether a group of veterans associated with the development of PTSD or not [123]. Moreover, a source level study by Bae and colleagues reported reduced source density of P300 in patients with PTSD when compared with healthy controls [4].

In addition, specific symptoms were related to different neural activity at different cortical regions.

Recently, a study that examined the source level functional connectivity of resting-state EEG reflected source level abnormalities in patient with PTSD [121]. In this study, theta and low beta frequency bands were significantly impaired in patients with PTSD compared with healthy control. As for the nodal clustering coefficients, that of the theta band negatively correlated with rumination and re-experience symptom scores, while that of the low beta band correlated positively with anxiety and pain severity. In theta frequency band, the right posterior cingulate cortex and left temporal cortex were significantly correlated with impact event scale of patients with PTSD. As evidenced in Fig. 7.7, brain anatomy-based information can be given in source-level EEG functional connectivity analysis. It could be a promising biomarker of PTSD as well as other psychiatric illness, as not only a disease classifier but also treatment response indicator of various psychotropic medication and cognitive behavioral treatment.

7.3.4.2 Obsessive Compulsive Disorder (OCD)

Patients with obsessive compulsive disorder (OCD) are characterized by inclinations to monitor their thoughts and behaviors with a desire to maintain control. Neurophysiological studies on OCD have focused on their tendency to monitor performance and patterns in processing errors. In particular, the error-related negativity (ERN), a neural alteration that occurs 50–100 ms after committing errors, has been of special interest. The ERN is thought to reflect activity of the response-monitoring system and greater magnitude ERNs are observed in patients with OCD [28, 51]. A major advantage of the ERN is that its enhancement in OCD seems to be independent of pharmacological or psychological interventions [23, 127]. In addition, similar patterns have been identified in subclinical OCD symptoms [99, 118], and non-affected first-degree relatives of patients with OCD, which suggest the possibility of a biological influence [13, 113].

Studies examining cortical activities in OCD have reported significant results as well. A study by Kopřivová and colleagues, who applied low-resolution electromagnetic tomography (LORETA) and independent component analysis (ICA), identified medial frontal hyperactivation in OCD [65]. In addition, patients with OCD were found to have greater relative power of lower alpha (8–10 Hz) frequency band in the left hemisphere at rest and during presentations of neutral, aversive, and OCD-related pictures. These changes in hemispheric alpha band has been proposed as biomarker for increased avoidance motivation in patients with OCD [45].

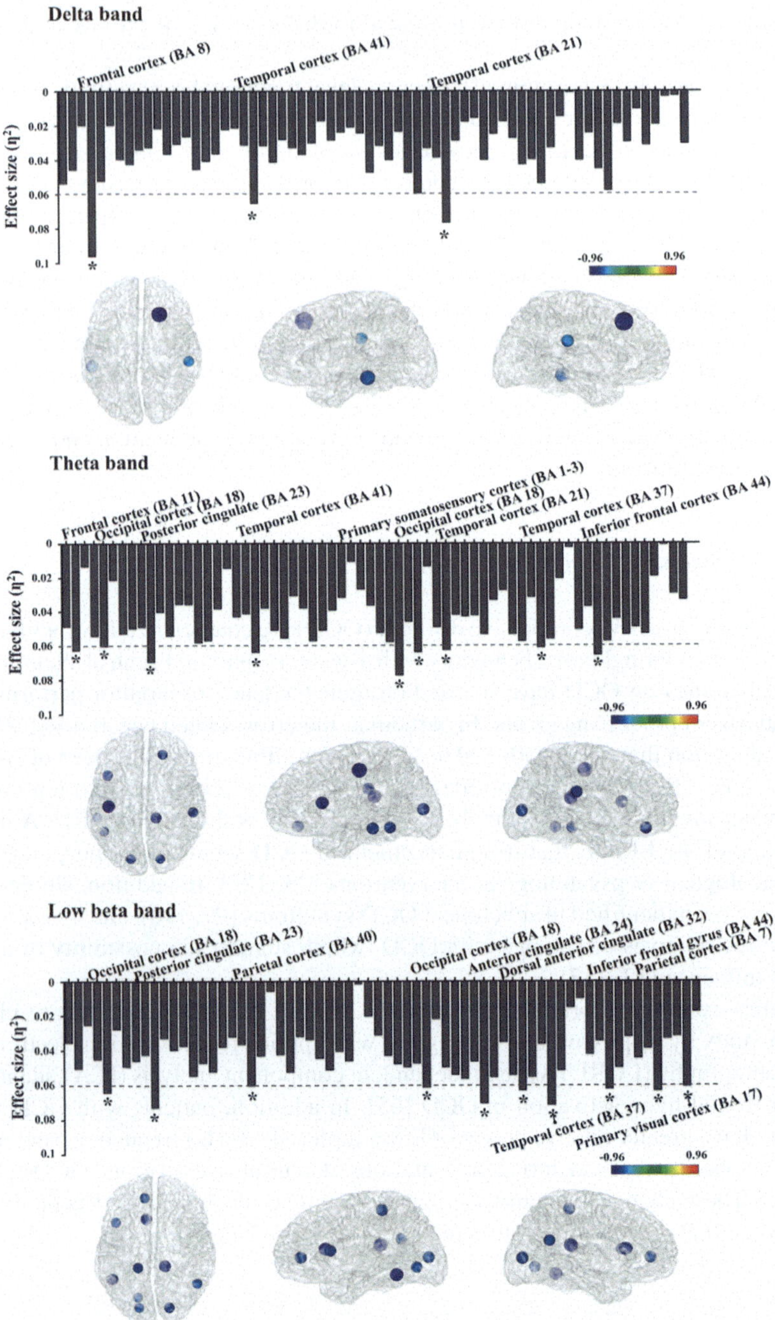

Delta band
Frontal cortex (BA 8)
Temporal cortex (BA 41)
Temporal cortex (BA 21)
Effect size (η²)
0
0.02
0.04
0.06
0.08
0.1
-0.96
0.96
Theta band
Frontal cortex (BA 11)
Occipital cortex (BA 18)
Posterior cingulate (BA 23)
Temporal cortex (BA 41)
Primary somatosensory cortex (BA 1-3)
Occipital cortex (BA 18)
Temporal cortex (BA 21)
Temporal cortex (BA 37)
Inferior frontal cortex (BA 44)
Effect size (η²)
0
0.02
0.04
0.06
0.08
0.1
-0.96
0.96
Low beta band
Occipital cortex (BA 18)
Posterior cingulate (BA 23)
Parietal cortex (BA 40)
Occipital cortex (BA 18)
Anterior cingulate (BA 24)
Dorsal anterior cingulate (BA 32)
Inferior frontal gyrus (BA 44)
Parietal cortex (BA 7)
Temporal cortex (BA 37)
Primary visual cortex (BA 17)
Effect size (η²)
0
0.02
0.04
0.06
0.08
0.1
-0.96
0.96

◂**Fig. 7.7** Effect-size of differences of nodal clustering coefficients between post-traumatic stress disorder (PTSD) and healthy controls in three frequency bands. The threshold value was set as 0.06 (medium effect). In the brain model, the density of colors and size of circles represent the difference direction and effect-size, respectively. *η240.06. Cited from the article of Shim et al. [121]

7.4 Diagnosis for Cognitive Impairment: Dementia and Minimal Cognitive Impairment

Neurophysiological markers, including EEG and ERP components, have been proposed in the literature to serve as indicators of cognitive decline [105]. This can further assist identification of current disease stage [3], and even treatment assessment [132]. Slowing of the EEG rhythm has been reported in dementia of the Alzheimer's (AD) type [83]. More specifically, theta band power has been found to increase and the alpha power to decrease in the left hemisphere for patients with AD. Simultaneously, all band powers showed decreased activity in the right hemisphere [1]. D'Onofrio and colleagues reported that the increase of delta and theta activity, and decrease of alpha distinguish normal aging, minimal cognitive impairment (MCI) and AD [17]. In a more recent study, Kim and colleagues reported that patients with AD with moderate severity had significantly increased theta and decreased beta power compared with healthy controls [61]. In patients with AD, theta power was significantly correlated with a poor performance for global cognition. However, beta power was positively correlated with a good performance for global cognition, attention, memory, visuospatial function, and executive function. Source activity analysis identified superior temporal gyrus, transverse temporal gyrus, insula, postcentral gyrus, cuneus, and lingual gyrus regions to be significantly affected in patients with AD.

It has been shown that MMN amplitude decreases with aging when short ISIs are used, while memory trace decays faster in patients with AD when long ISIs are applied [110]. The MMN has also been used as a feature of prognosis in MCI progression to AD [105].

As previously mentioned, P300 is one of the most studied ERP components with regards to attention, working memory, and cognitive decline. Therefore, it has been often studied in respect to dementia and cognitive impairment. The P3a has been associated with attention mechanisms and novel stimulus processing, while P3b, the second subcomponent, has been more related to stimulus evaluation and decision making [111]. The latency and amplitude of P300 have been shown to aptly reflect neurodegeneration progress [67, 89], and significant differences in its characteristics have been found between controls and MCI or AD patients [8, 27, 44, 105, 109]. However, several studies have reported minor variability in P300 features as well [2, 70], which raises questions to whether the P300 can act as a reliable biomarker of cognitive impairment.

Park et al. [107] computed the global field synchronization (GFS), a measure of functional synchronization, which were lower in beta1, beta2, beta3, and full bands in AD patients than in healthy controls. These values positively correlated with the mini mental state examination (MMSE) and clinical dementia rating (CDR) scores [107]. Lee and colleagues inspected the global synchronization index (GSI), which quantifies synchronization between neuronal signals at multiple sites [71]. They found that in patients with AD, GSI values were negatively correlated with MMSE scores in the delta bands, but positively correlated in the beta1 and gamma band. In addition, GSI values were positively correlated with CDR scores in the delta bands, but negatively correlated in the gamma band. The EEG measurements could be promising biomarker of dementia. However, tight artifact removal and data collection criteria should precede the measurements for standardized settings for researchers working in various experimental situations.

7.5 Conclusion

In this chapter, we reviewed possible EEG biomarkers of major psychiatric disorders. Among the various EEG characteristics, P300, functional connectivity network, LDAEP, ASSR, MMN, and alpha asymmetry are promising EEG biomarkers that demand more attention in the future. In concordance with the development of computational analytic methods, source level functional connectivity network could make a new breakthrough for EEG based diagnosis of various neuropsychiatric disorder, namely in schizophrenia, PTSD, major depressive disorder, and dementia of various types. In patients with major depressive disorder, alpha asymmetry could be a promising biomarker. However, more research should be done to uncover the confounding factor of the core pathophysiology of major depressive disorders.

Sensor-level analysis could be contaminated with volume conduction and various movement artifacts. However, source-level analysis could be used relatively free of those artifacts, even though there are some technical burden compared to sensor-level analysis. Importantly, information about brain regional abnormality could be gained from these source-level analyses, and this information could be a big step to discover diagnostic and prognosis biomarkers of various neuropsychiatric disorders.

Acknowledgments This research was supported by the Brain Research Program through the National Research Foundation of Korea (NRF), funded by the Ministry of Science, ICT & Future Planning (NRF-2015M3C7A1028252), the Korea Science and Engineering Foundation (KOSEF), funded by the Korean government (NRF-2018R1A2A2A05018505), and by the 2017 creative research program of Inje University.

References

1. H. Aghajani, E. Zahedi, M. Jalili, A. Keikhosravi, B.V. Vahdat, Diagnosis of early Alzheimer's disease based on EEG source localization and a standardized realistic head model. IEEE J. Biomed. Health Inform. **17**(6), 1039–45 (2013)
2. J.W. Ashford, K.L. Coburn, T.L. Rose, P.J. Bayley, P300 energy loss in aging and Alzheimer's disease. J. Alzheimers Dis. **26**(s3), 229–38 (2011)

3. C. Babiloni, R. Lizio, C. Del Percio, N. Marzano, A. Soricelli, E. Salvatore, R. Ferri, F.I. Cosentino, G. Tedeschi, P. Montella, Cortical sources of resting state EEG rhythms are sensitive to the progression of early stage Alzheimer's disease. J. Alzheimers Dis. **34**(4), 1015–35 (2013)
4. K.-Y. Bae, D.-W. Kim, C.-H. Im, S.-H. Lee, Source imaging of P300 auditory evoked potentials and clinical correlations in patients with posttraumatic stress disorder. Prog. Neuropsychopharmacol. Biol. Psychiatry **35**(8), 1908–17 (2011)
5. B. Bandelow, D. Baldwin, M. Abelli, B. Bolea-Alamanac, M. Bourin, S.R. Chamberlain, E. Cinosi, S. Davies, K. Domschke, N. Fineberg, Biological markers for anxiety disorders, OCD and PTSD: a consensus statement. Part II: Neurochemistry, neurophysiology and neurocognition. World J. Biol. Psychiatry **18**(3), 162–214 (2017)
6. A. Bari, T.W. Robbins, Inhibition and impulsivity: behavioral and neural basis of response control. Prog. Neurobiol. **108**, 44–79 (2013)
7. P. Barttfeld, A. Petroni, S. Báez, H. Urquina, M. Sigman, M. Cetkovich, T. Torralva, F. Torrente, A. Lischinsky, X. Castellanos, Functional connectivity and temporal variability of brain connections in adults with attention deficit/hyperactivity disorder and bipolar disorder. Neuropsychobiology **69**(2), 65–75 (2014)
8. K. Bennys, G. Rondouin, E. Benattar, A. Gabelle, J. Touchon, Can event-related potential predict the progression of mild cognitive impairment? J. Clin. Neurophysiol. **28**(6), 625–32 (2011)
9. J. Bhattacharya, Reduced degree of long-range phase synchrony in pathological human brain. Acta Neurobiol. Exp. **61**(4), 309–18 (2001)
10. M. Bolanos, E.M. Bernat, B. He, S. Aviyente, A weighted small world network measure for assessing functional connectivity. J. Neurosci. Methods **212**(1), 133–42 (2013)
11. P. Brambilla, M. Bellani, P.-H. Yeh, J.C. Soares, M. Tansella, White matter connectivity in bipolar disorder. Int. Rev. Psychiatry **21**(4), 380–6 (2009)
12. E. Bullmore, O. Sporns, Complex brain networks: graph theoretical analysis of structural and functional systems. Nat. Rev. Neurosci. **10**(3), 186 (2009)
13. M. Carrasco, S.M. Harbin, J.K. Nienhuis, K.D. Fitzgerald, W.J. Gehring, G.L. Hanna, Increased error-related brain activity in youth with obsessive-compulsive disorder and unaffected siblings. Depress. Anxiety **30**(1), 39–46 (2013)
14. S.V. Catts, A.-M. Shelley, P.B. Ward, B. Liebert, Brain potential evidence for an auditory sensory memory deficit in schizophrenia. Am. J. Psychiatry **152**(2), 213 (1995)
15. H.W. Chase, M.L. Phillips, Elucidating neural network functional connectivity abnormalities in bipolar disorder: toward a harmonized methodological approach. Biol. Psychiatry Cogn. Neurosci. Neuroimaging**1**(3), 288–98 (2016)
16. J.A. Coan, J.J. Allen, Frontal EEG asymmetry and the behavioral activation and inhibition systems. Psychophysiology **40**(1), 106–14 (2003)
17. F. d'Onofrio, S. Salvia, V. Petretta, V. Bonavita, G. Rodriguez, G. Tedeschi, Quantified-EEG in normal aging and dementias. Acta Neurol. Scand. **93**(5), 336–45 (1996)
18. R.J. Davidson, EEG measures of cerebral asymmetry: conceptual and methodological issues. Int. J. Neurosci. **39**(1–2), 71–89 (1988)
19. R.J. Davidson, *Emotion and Affective Style: Hemispheric Substrates* (SAGE Publications, Sage, Los Angeles, CA, 1992)
20. S. Debener, A. Beauducel, D. Nessler, B. Brocke, H. Heilemann, J. Kayser, Is resting anterior EEG alpha asymmetry a trait marker for depression? Neuropsychobiology **41**(1), 31–7 (2000)
21. M.A. Diego, T. Field, M. Hernandez-Reif, BIS/BAS scores are correlated with frontal EEG asymmetry in intrusive and withdrawn depressed mothers. Infant. Mental Health J. **22**(6), 665–75 (2001)
22. R. Ehrlichman, M. Gandal, C. Maxwell, M. Lazarewicz, L. Finkel, D. Contreras, B. Turetsky, S. Siegel, *N*-methyl-D-aspartic acid receptor antagonist–induced frequency oscillations in mice recreate pattern of electrophysiological deficits in schizophrenia. Neuroscience **158**(2), 705–12 (2009)

23. T. Endrass, B. Schuermann, C. Kaufmann, R. Spielberg, R. Kniesche, N. Kathmann, Performance monitoring and error significance in patients with obsessive-compulsive disorder. Biol. Psychol. **84**(2), 257–63 (2010)
24. F.D.V. Fallani, A. Maglione, F. Babiloni, D. Mattia, L. Astolfi, G. Vecchiato, A. De Rinaldis, S. Salinari, E. Pachou, S. Micheloyannis, Cortical network analysis in patients affected by schizophrenia. Brain Topogr. **23**(2), 214–20 (2010)
25. A.A. Fingelkurts, A.A. Fingelkurts, Altered structure of dynamic electroencephalogram oscillatory pattern in major depression. Biol. Psychiat. **77**(12), 1050–60 (2015)
26. A.A. Fingelkurts, A.A. Fingelkurts, H. Rytsälä, K. Suominen, E. Isometsä, S. Kähkönen, Impaired functional connectivity at EEG alpha and theta frequency bands in major depression. Hum. Brain Mapp. **28**(3), 247–61 (2007)
27. S.T. Fix, J.E. Arruda, F. Andrasik, J. Beach, K. Groom, Using visual evoked potentials for the early detection of amnestic mild cognitive impairment: a pilot investigation. Int. J. Geriatr. Psychiatry **30**(1), 72–9 (2015)
28. W.J. Gehring, B. Goss, M.G. Coles, D.E. Meyer, E. Donchin, A neural system for error detection and compensation. Psychol. Sci. **4**(6), 385–90 (1993)
29. G. Gonzalez-Burgos, D.A. Lewis, NMDA receptor hypofunction, parvalbumin-positive neurons, and cortical gamma oscillations in schizophrenia. Schizophr. Bull. **38**(5), 950–7 (2012)
30. E. Gordon, D.M. Palmer, N. Cooper, EEG alpha asymmetry in schizophrenia, depression, PTSD, panic disorder, ADHD and conduct disorder. Clin. EEG Neurosci. **41**(4), 178–83 (2010)
31. I.H. Gotlib, EEG alpha asymmetry, depression, and cognitive functioning. Cogn. Emot. **12**(3), 449–78 (1998)
32. J.A. Gray, The psychophysiological basis of introversion-extraversion. Behav. Res. Ther. **8**(3), 249–66 (1970)
33. S. Gudmundsson, T.P. Runarsson, S. Sigurdsson, G. Eiriksdottir, K. Johnsen, Reliability of quantitative EEG features. Clin. Neurophysiol. **118**(10), 2162–71 (2007)
34. C.N. Gupta, V.D. Calhoun, S. Rachakonda, J. Chen, V. Patel, J. Liu, J. Segall, B. Franke, M.P. Zwiers, A. Arias-Vasquez, Patterns of gray matter abnormalities in schizophrenia based on an international mega-analysis. Schizophr. Bull. **41**(5), 1133–42 (2014)
35. J.P. Hamm, C.S. Gilmore, B.A. Clementz, Augmented gamma band auditory steady-state responses: support for NMDA hypofunction in schizophrenia. Schizophr. Res. **138**(1), 1–7 (2012)
36. E. Harmon-Jones, J.J. Allen, Behavioral activation sensitivity and resting frontal EEG asymmetry: covariation of putative indicators related to risk for mood disorders. J. Abnorm. Psychol. **106**(1), 159 (1997)
37. S. Haufe, V.V. Nikulin, K.-R. Müller, G. Nolte, A critical assessment of connectivity measures for EEG data: a simulation study. Neuroimage **64**, 120–33 (2013)
38. U. Hegerl, J. Gallinat, G. Juckel, Event-related potentials: do they reflect central serotonergic neurotransmission and do they predict clinical response to serotonin agonists? J. Affect. Disord. **62**(1), 93–100 (2001)
39. J.B. Henriques, R.J. Davidson, Left frontal hypoactivation in depression. J. Abnorm. Psychol. **100**(4), 535 (1991)
40. A.T. Herdman, O. Lins, P. Van Roon, D.R. Stapells, M. Scherg, T.W. Picton, Intracerebral sources of human auditory steady-state responses. Brain Topogr. **15**(2), 69–86 (2002)
41. J. Higley, M. Linnoila, Low central nervous system serotonergic activity is traitlike and correlates with impulsive behavior. Ann. N. Y. Acad. Sci. **836**(1), 39–56 (1997)
42. Y. Hirano, N. Oribe, S. Kanba, T. Onitsuka, P.G. Nestor, K.M. Spencer, Spontaneous gamma activity in schizophrenia. JAMA Psychiatry **72**(8), 813–21 (2015)
43. L.E. Hong, A. Summerfelt, R. McMahon, H. Adami, G. Francis, A. Elliott, R.W. Buchanan, G.K. Thaker, Evoked gamma band synchronization and the liability for schizophrenia. Schizophr. Res. **70**(2), 293–302 (2004)
44. A.S. Howe, Meta-analysis of the endogenous N200 latency event-related potential subcomponent in patients with Alzheimer's disease and mild cognitive impairment. Clin. Neurophysiol. **125**(6), 1145–51 (2014)

45. M. Ischebeck, T. Endrass, D. Simon, N. Kathmann, Altered frontal EEG asymmetry in obsessive-compulsive disorder. Psychophysiology **51**(7), 596–601 (2014)
46. M.P. Jadi, M.M. Behrens, T.J. Sejnowski, Abnormal gamma oscillations in N-methyl-D-aspartate receptor hypofunction models of schizophrenia. Biol. Psychiat. **79**(9), 716–26 (2016)
47. M. Jalili, M.G. Knyazeva, EEG-based functional networks in schizophrenia. Comput. Biol. Med. **41**(12), 1178–86 (2011)
48. D.C. Javitt, M. Steinschneider, C.E. Schroeder, J.C. Arezzo, Role of cortical *N*-methyl-D-aspartate receptors in auditory sensory memory and mismatch negativity generation: implications for schizophrenia. Proc. Natl. Acad. Sci. **93**(21), 11962–7 (1996)
49. N. Jaworska, P. Blier, W. Fusee, V. Knott, Alpha power, alpha asymmetry and anterior cingulate cortex activity in depressed males and females. J. Psychiatr. Res. **46**(11), 1483–91 (2012)
50. H.-G. Jeong, Y.-H. Ko, C. Han, Y.-K. Kim, S.-H. Joe, Distinguishing quantitative electroencephalogram findings between adjustment disorder and major depressive disorder. Psychiatry Investig. **10**(1), 62–8 (2013)
51. S. Johannes, B.M. Wieringa, W. Nager, D. Rada, R. Dengler, H.M. Emrich, T.F. Münte, D.E. Dietrich, Discrepant target detection and action monitoring in obsessive–compulsive disorder. Psychiatry Res. Neuroimaging **108**(2), 101–10 (2001)
52. G. Juckel, P. Mavrogiorgou, S. Bredemeier, J. Gallinat, T. Frodl, C. Schulz, H.J. Moller, U. Hegerl, Loudness dependence of primary auditory-cortex-evoked activity as predictor of therapeutic outcome to prophylactic lithium treatment in affective disorders—a retrospective study. Pharmacopsychiatry **37**(2), 46–51 (2004)
53. G. Juckel, M. Molnár, U. Hegerl, V. Csépe, G. Karmos, Auditory-evoked potentials as indicator of brain serotonergic activity first evidence in behaving cats. Biol. Psychiat. **41**(12), 1181–95 (1997)
54. J.W. Kam, A.R. Bolbecker, B.F. O'Donnell, W.P. Hetrick, C.A. Brenner, Resting state EEG power and coherence abnormalities in bipolar disorder and schizophrenia. J. Psychiatr. Res. **47**(12), 1893–901 (2013)
55. J.T. Kantrowitz, D.C. Javitt, *N*-methyl-D-aspartate (NMDA) receptor dysfunction or dysregulation: the final common pathway on the road to schizophrenia? Brain Res. Bull. **83**(3–4), 108–21 (2010)
56. K.L. Kasch, J. Rottenberg, B.A. Arnow, I.H. Gotlib, Behavioral activation and inhibition systems and the severity and course of depression. J. Abnorm. Psychol. **111**(4), 589 (2002)
57. Y. Kawakubo, K. Kasai, Support for an association between mismatch negativity and social functioning in schizophrenia. Prog. Neuropsychopharmacol. Biol. Psychiatry **30**(7), 1367–8 (2006)
58. A. Kemp, K. Griffiths, K. Felmingham, S. Shankman, W. Drinkenburg, M. Arns, C. Clark, R. Bryant, Disorder specificity despite comorbidity: resting EEG alpha asymmetry in major depressive disorder and post-traumatic stress disorder. Biol. Psychol. **85**(2), 350–4 (2010)
59. M. Kiang, G.A. Light, J. Prugh, S. Coulson, D.L. Braff, M. Kutas, Cognitive, neurophysiological, and functional correlates of proverb interpretation abnormalities in schizophrenia. J. Int. Neuropsychol. Soc. **13**(4), 653–63 (2007)
60. D.-J. Kim, A.R. Bolbecker, J. Howell, O. Rass, O. Sporns, W.P. Hetrick, A. Breier, B.F. O'Donnell, Disturbed resting state EEG synchronization in bipolar disorder: a graph-theoretic analysis. NeuroImage: Clinical **2**, 414–23 (2013)
61. J.-S. Kim, S.-H. Lee, G. Park, S. Kim, S.-M. Bae, D.-W. Kim, C.-H. Im, Clinical implications of quantitative electroencephalography and current source density in patients with Alzheimer's disease. Brain Topogr. **25**(4), 461–74 (2012)
62. J.S. Kim, S. Kim, W. Jung, C.-H. Im, S.-H. Lee, Auditory evoked potential could reflect emotional sensitivity and impulsivity. Sci. Rep. **6**, 37683 (2016)
63. S. Kim, S.-K. Jang, D.-W. Kim, M. Shim, Y.-W. Kim, C.-H., S.-H., *Auditory-steady-state responses and cortical volume in patients with schizophrenia* (unpublished data)
64. V. Knott, C. Mahoney, S. Kennedy, K. Evans, EEG power, frequency, asymmetry and coherence in male depression. Psychiatry Res. Neuroimaging **106**(2), 123–40 (2001)

65. J. Kopřivová, M. Congedo, J. Horáček, J. Praško, M. Raszka, M. Brunovský, B. Kohútová, C. Höschl, EEG source analysis in obsessive–compulsive disorder. Clin. Neurophysiol. **122**(9), 1735–43 (2011)
66. T. Korotkova, E.C. Fuchs, A. Ponomarenko, J. von Engelhardt, H. Monyer, NMDA receptor ablation on parvalbumin-positive interneurons impairs hippocampal synchrony, spatial representations, and working memory. Neuron **68**(3), 557–69 (2010)
67. C.-L. Lai, R.-T. Lin, L.-M. Liou, C.-K. Liu, The role of event-related potentials in cognitive decline in Alzheimer's disease. Clin. Neurophysiol. **121**(2), 194–9 (2010)
68. M. Leboyer, D.J. Kupfer, Bipolar disorder: new perspectives in health care and prevention. J. Clin. Psychiatry **71**(12), 1689 (2010)
69. B.-H. Lee, Y.-M. Park, S.-H. Lee, M. Shim, Prediction of long-term treatment response to selective serotonin reuptake inhibitors (SSRIs) using scalp and source loudness dependence of auditory evoked potentials (LDAEP) analysis in patients with major depressive disorder. Int. J. Mol. Sci. **16**(3), 6251–65 (2015)
70. M.-S. Lee, S.-H. Lee, E.-O. Moon, Y.-J. Moon, S. Kim, S.-H. Kim, I.-K. Jung, Neuropsychological correlates of the P300 in patients with Alzheimer's disease. Prog. Neuropsychopharmacol. Biol. Psychiatry **40**, 62–9 (2013)
71. S.-H. Lee, Y.-M. Park, D.-W. Kim, C.-H. Im, Global synchronization index as a biological correlate of cognitive decline in Alzheimer's disease. Neurosci. Res. **66**(4), 333–9 (2010)
72. S.-H. Lee, S. Yoon, J.-I. Kim, S.-H. Jin, C.K. Chung, Functional connectivity of resting state EEG and symptom severity in patients with post-traumatic stress disorder. Prog. Neuropsychopharmacol. Biol. Psychiatry **51**, 51–7 (2014)
73. S.H. Lee, J.H. Kim, J.H. Lee, S. Kim, Y.M. Park, S.M. Bae, J.S. Choo, Aberrant response of selective serotonin reuptake inhibitor in two patients with high N100 amplitude slope. Korean J. Psychopharmacol **19**, 341–7 (2008)
74. S.H. Lee, K. Sung, K.S. Lee, E. Moon, C.G. Kim, Mismatch negativity is a stronger indicator of functional outcomes than neurocognition or theory of mind in patients with schizophrenia. Prog. Neuropsychopharmacol. Biol. Psychiatry **48**, 213–9 (2014)
75. T.-W. Lee, Y.W.-Y. Yu, M.-C. Chen, T.-J. Chen, Cortical mechanisms of the symptomatology in major depressive disorder: a resting EEG study. J. Affect. Disord. **131**(1), 243–50 (2011)
76. A.F. Leuchter, I.A. Cook, A.M. Hunter, C. Cai, S. Horvath, Resting-state quantitative electroencephalography reveals increased neurophysiologic connectivity in depression. PLoS ONE **7**(2), e32508 (2012)
77. D.A. Lewis, A.A. Curley, J.R. Glausier, D.W. Volk, Cortical parvalbumin interneurons and cognitive dysfunction in schizophrenia. Trends Neurosci. **35**(1), 57–67 (2012)
78. A.L. Lieber, Diagnosis and subtyping of depressive disorders by quantitative electroencephalography: II. Interhemispheric measures are abnormal in major depressives and frequency analysis may discriminate certain subtypes. Hillside J. Clin. Psychiatry (1988)
79. G.A. Light, D.L. Braff, Mismatch negativity deficits are associated with poor functioning in schizophrenia patients. Arch. Gen. Psychiatry **62**(2), 127–36 (2005a)
80. G.A. Light, D.L. Braff, Stability of mismatch negativity deficits and their relationship to functional impairments in chronic schizophrenia. Am. J. Psychiatry **162**(9), 1741–3 (2005b)
81. G.A. Light, N.R. Swerdlow, D.L. Braff, Preattentive sensory processing as indexed by the MMN and P3a brain responses is associated with cognitive and psychosocial functioning in healthy adults. J. Cogn. Neurosci. **19**(10), 1624–32 (2007)
82. I. Lobo, L.C. Portugal, I. Figueira, E. Volchan, I. David, M.G. Pereira, L. de Oliveira, EEG correlates of the severity of posttraumatic stress symptoms: a systematic review of the dimensional PTSD literature. J. Affect. Disord. **183**, 210–20 (2015)
83. C. Logar, W. Grabmair, G. Schneider, H. Lechner, EEG changes in senile dementia of the Alzheimer type. EEG-EMG Zeitschrift fur Elektroenzephalographie, Elektromyographie und verwandte Gebiete **18**(4), 214–7 (1987)
84. S.J. Luck (2005) Ten simple rules for designing ERP experiments. *Event-related Potentials: A Methods Handbook 262083337*

85. S.B. Manuck, J.D. Flory, J.M. McCaffery, K.A. Matthews, J.J. Mann, M.F. Muldoon, Aggression, impulsivity, and central nervous system serotonergic responsivity in a nonpatient sample. Neuropsychopharmacology **19**(4), 287–99 (1998)
86. D.H. Mathalon, V.S. Sohal, Neural oscillations and synchrony in brain dysfunction and neuropsychiatric disorders: it's about time. JAMA Psychiatry **72**(8), 840–4 (2015)
87. K. Matsuo, M. Nicoletti, K. Nemoto, J.P. Hatch, M.A. Peluso, F.G. Nery, J.C. Soares, A voxel-based morphometry study of frontal gray matter correlates of impulsivity. Hum. Brain Mapp. **30**(4), 1188–95 (2009)
88. J.M. McNally, R.W. McCarley, Gamma band oscillations: a key to understanding schizophrenia symptoms and neural circuit abnormalities. Curr. Opin. Psychiatry **29**(3), 202 (2016)
89. S. Medvidovic, M. Titlic, M. Maras-Simunic, P300 evoked potential in patients with mild cognitive impairment. Acta Informatica Med. **21**(2), 89 (2013)
90. T. Meyer, T. Smeets, T. Giesbrecht, C.W. Quaedflieg, F.T. Smulders, E.H. Meijer, H.L. Merckelbach, The role of frontal EEG asymmetry in post-traumatic stress disorder. Biol. Psychol. **108**, 62–77 (2015)
91. S. Micheloyannis, E. Pachou, C.J. Stam, M. Breakspear, P. Bitsios, M. Vourkas, S. Erimaki, M. Zervakis, Small-world networks and disturbed functional connectivity in schizophrenia. Schizophr. Res. **87**(1), 60–6 (2006)
92. E. Monkul, J.P. Hatch, M.A. Nicoletti, S. Spence, P. Brambilla, ALTd Lacerda, R.B. Sassi, A. Mallinger, M. Keshavan, J.C. Soares, Fronto-limbic brain structures in suicidal and non-suicidal female patients with major depressive disorder. Mol. Psychiatry **12**(4), 360 (2007)
93. T. Montez, K. Linkenkaer-Hansen, B. Van Dijk, C. Stam, Synchronization likelihood with explicit time-frequency priors. Neuroimage **33**(4), 1117–25 (2006)
94. R. Näätänen, A.W. Gaillard, S. Mäntysalo, Early selective-attention effect on evoked potential reinterpreted. Acta Physiol. (Oxf) **42**(4), 313–29 (1978)
95. R. Näätänen, T. Kujala, C. Escera, T. Baldeweg, K. Kreegipuu, S. Carlson, C. Ponton, The mismatch negativity (MMN)—a unique window to disturbed central auditory processing in ageing and different clinical conditions. Clin. Neurophysiol. **123**(3), 424–58 (2012)
96. K. O'connor, J. Shaw, C. Ongley, The EEG and differential diagnosis in psychogeriatrics. Br. J. Psychiatry **135**(2), 156–62 (1979)
97. B.V. O'neill, R.J. Croft, P.J. Nathan, The loudness dependence of the auditory evoked potential (LDAEP) as an in vivo biomarker of central serotonergic function in humans: rationale, evaluation and review of findings. Human Psychopharmacol: Clin. Exp. **23**(5), 355–70 (2008)
98. B.F. O'Donnell, J.L. Vohs, G.P. Krishnan, O. Rass, W.P. Hetrick, S.L. Morzorati, The auditory steady-state response (ASSR): a translational biomarker for schizophrenia. Suppl. Clin. Neurophysiol. **62**, 101 (2013)
99. S.A. O'Toole, M. Weinborn, A.M. Fox, Performance monitoring among non-patients with obsessive–compulsive symptoms: ERP evidence of aberrant feedback monitoring. Biol. Psychol. **91**(2), 221–8 (2012)
100. Y. Oda, T. Onitsuka, R. Tsuchimoto, S. Hirano, N. Oribe, T. Ueno, Y. Hirano, I. Nakamura, T. Miura, S. Kanba, Gamma band neural synchronization deficits for auditory steady state responses in bipolar disorder patients. PLoS ONE **7**(7), e39955 (2012)
101. O. Oluboka, S. Stewart, V. Sharma, D. Mazmanian, E. Persad, Preliminary assessment of intrahemispheric QEEG measures in bipolar mood disorders. Can. J. Psychiatry **47**(4), 368–74 (2002)
102. A. Özerdem, B. Güntekin, İ. Atagün, E. Başar, Brain oscillations in bipolar disorder in search of new biomarkers. Suppl. Clin. Neurophysiol. **62**, 207–21 (2013)
103. A. Özerdem, B. Güntekin, İ. Atagün, B. Turp, E. Başar, Reduced long distance gamma (28–48 Hz) coherence in euthymic patients with bipolar disorder. J. Affect. Disord. **132**(3), 325–32 (2011)
104. A. Özerdem, B. Güntekin, E. Saatçi, Z. Tunca, E. Başar, Disturbance in long distance gamma coherence in bipolar disorder. Prog. Neuropsychopharmacol. Biol. Psychiatry **34**(6), 861–5 (2010)

105. V. Papaliagkas, V. Kimiskidis, M. Tsolaki, G. Anogianakis, Cognitive event-related potentials: longitudinal changes in mild cognitive impairment. Clin. Neurophysiol. **122**(7), 1322–6 (2011)
106. C.A. Park, R.-J. Kwon, S. Kim, H.-R. Jang, J.-H. Chae, T. Kim, J. Jeong (ed.) Decreased phase synchronization of the EEG in patients with major depressive disorder, in *World Congress on Medical Physics and Biomedical Engineering 2006* (Springer, Berlin, 2007)
107. Y.-M. Park, H.-J. Che, C.-H. Im, H.-T. Jung, S.-M. Bae, S.-H. Lee, Decreased EEG synchronization and its correlation with symptom severity in Alzheimer's disease. Neurosci. Res. **62**(2), 112–7 (2008)
108. Y.-M. Park, D.-W. Kim, S. Kim, C.-H. Im, S.-H. Lee, The loudness dependence of the auditory evoked potential (LDAEP) as a predictor of the response to escitalopram in patients with generalized anxiety disorder. Psychopharmacology **213**(2–3), 625–32 (2011)
109. R.V. Pedroso, F.J. Fraga, D.I. Corazza, C.A.A. Andreatto, F.G. de Melo Coelho, J.L.R. Costa, R.F. Santos-Galduróz, P300 latency and amplitude in Alzheimer's disease: a systematic review. Braz. J. Otorhinolaryngol. **78**(4), 126–32 (2012)
110. E. Pekkonen, Mismatch negativity in aging and in Alzheimer's and Parkinson's diseases. Audiol. Neurotol. **5**(3–4), 216–24 (2000)
111. J. Polich, Updating P300: an integrative theory of P3a and P3b. Clin. Neurophysiol. **118**(10), 2128–48 (2007)
112. O. Rass, G. Krishnan, C.A. Brenner, W.P. Hetrick, C.C. Merrill, A. Shekhar, B.F. O'Donnell, Auditory steady state response in bipolar disorder: relation to clinical state, cognitive performance, medication status, and substance disorders. Bipolar Disord. **12**(8), 793–803 (2010)
113. A. Riesel, T. Endrass, C. Kaufmann, N. Kathmann, Overactive error-related brain activity as a candidate endophenotype for obsessive-compulsive disorder: evidence from unaffected first-degree relatives. Am. J. Psychiatry **168**(3), 317–24 (2011)
114. M. Rubinov, S.A. Knock, C.J. Stam, S. Micheloyannis, A.W. Harris, L.M. Williams, M. Breakspear, Small-world properties of nonlinear brain activity in schizophrenia. Hum. Brain Mapp. **30**(2), 403–16 (2009)
115. M. Rubinov, O. Sporns, Complex network measures of brain connectivity: uses and interpretations. Neuroimage **52**(3), 1059–69 (2010)
116. E. Salinas, T.J. Sejnowski, Correlated neuronal activity and the flow of neural information. Nat. Rev. Neurosci. **2**(8), 539 (2001)
117. D.F. Salisbury, M.E. Shenton, C.B. Griggs, A. Bonner-Jackson, R.W. McCarley, Mismatch negativity in chronic schizophrenia and first-episode schizophrenia. Arch. Gen. Psychiatry **59**(8), 686–94 (2002)
118. D.L. Santesso, S.J. Segalowitz, L.A. Schmidt, Error-related electrocortical responses are enhanced in children with obsessive–compulsive behaviors. Dev. Neuropsychol. **29**(3), 431–45 (2006)
119. S. Schinkel, G. Zamora-López, O. Dimigen, W. Sommer, J. Kurths, Functional network analysis reveals differences in the semantic priming task. J. Neurosci. Methods **197**(2), 333–9 (2011)
120. A.E. Şevik, A.E.A. Yağcıoğlu, S. Yağcıoğlu, S. Karahan, N. Gürses, M. Yıldız, Neuropsychological performance and auditory event related potentials in schizophrenia patients and their siblings: a family study. Schizophr. Res. **130**(1), 195–202 (2011)
121. M. Shim, C. Im, S. Lee, Disrupted cortical brain network in post-traumatic stress disorder patients: a resting-state electroencephalographic study. Transl. Psychiatry **7**(9), e1231 (2017)
122. M. Shim, D.W. Kim, S.H. Lee, C.H. Im, Disruptions in small-world cortical functional connectivity network during an auditory oddball paradigm task in patients with schizophrenia. Schizophr. Res. **156**(2–3), 197–203 (2014)
123. I.-W. Shu, J.A. Onton, R.M. O'Connell, A.N. Simmons, S.C. Matthews, Combat veterans with comorbid PTSD and mild TBI exhibit a greater inhibitory processing ERP from the dorsal anterior cingulate cortex. Psychiatry Res. Neuroimaging **224**(1), 58–66 (2014)
124. D.R. Simkin, R.W. Thatcher, J. Lubar, Quantitative EEG and neurofeedback in children and adolescents. Child Adolesc Psychiatr Clin **23**(3), 427–64 (2014)

125. C.J. Stam, J.C. Reijneveld, Graph theoretical analysis of complex networks in the brain. Nonlinear Biomed Physics **1**(1), 3 (2007)
126. D.J. Stein, E. Hollander, M.R. Liebowitz (1993) Neurobiology of impulsivity and the impulse control disorders. J. Neuropsychiatry Clin Neurosci.
127. E.R. Stern, Y. Liu, W.J. Gehring, J.J. Lister, G. Yin, J. Zhang, K.D. Fitzgerald, J.A. Himle, J.L. Abelson, S.F. Taylor, Chronic medication does not affect hyperactive error responses in obsessive-compulsive disorder. Psychophysiology **47**(5), 913–20 (2010)
128. Y. Sun, Y. Li, Y. Zhu, X. Chen, S. Tong, Electroencephalographic differences between depressed and control subjects: an aspect of interdependence analysis. Brain Res. Bull. **76**(6), 559–64 (2008)
129. S.K. Sutton, R.J. Davidson, Prefrontal brain asymmetry: A biological substrate of the behavioral approach and inhibition systems. Psychol. Sci. **8**(3), 204–10 (1997)
130. H. Thuné, M. Recasens, P.J. Uhlhaas, The 40-Hz auditory steady-state response in patients with schizophrenia: a meta-analysis. JAMA Psychiatry **73**(11), 1145–53 (2016)
131. H. Tiitinen, P. May, R. Näätänen, The transient 40-Hz response, mismatch negativity, and attentional processes in humans. Prog. Neuropsychopharmacol. Biol. Psychiatry **21**(5), 751–71 (1997)
132. A. Tsolaki, D. Kazis, I. Kompatsiaris, V. Kosmidou, M. Tsolaki (2014) Electroencephalogram and Alzheimer's disease: clinical and research approaches. Int. J. Alzheimer's Dis. (2014)
133. P.J. Uhlhaas, W. Singer, Abnormal neural oscillations and synchrony in schizophrenia. Nat. Rev. Neurosci. **11**(2), 100 (2010)
134. D. Umbricht, J.A. Bates, J.A. Lieberman, J.M. Kane, D.C. Javitt, Electrophysiological indices of automatic and controlled auditory information processing in first-episode, recent-onset and chronic schizophrenia. Biol. Psychiat. **59**(8), 762–72 (2006)
135. D. Umbricht, D. Javitt, G. Novak, J. Bates, S. Pollack, J. Lieberman, J. Kane, Effects of clozapine on auditory event-related potentials in schizophrenia. Biol. Psychiat. **44**(8), 716–25 (1998)
136. D. Umbricht, D. Javitt, G. Novak, J. Bates, S. Pollack, J. Lieberman, J. Kane, Effects of risperidone on auditory event-related potentials in schizophrenia. Int. J. Neuropsychopharmacol. **2**(4), 299–304 (1999)
137. D. Umbricht, R. Koller, F.X. Vollenweider, L. Schmid, Mismatch negativity predicts psychotic experiences induced by NMDA receptor antagonist in healthy volunteers. Biol. Psychiat. **51**(5), 400–6 (2002)
138. B. Velasques, J. Bittencourt, C. Diniz, S. Teixeira, L.F. Basile, J.I. Salles, F. Novis, L.A. Silveira, RdA da Silva, Teixeira A. de Lima, Changes in saccadic eye movement (SEM) and quantitative EEG parameter in bipolar patients. J. Affect. Disord. **145**(3), 378–85 (2013)
139. J. Volkert, H. Schulz, M. Härter, O. Wlodarczyk, S. Andreas, The prevalence of mental disorders in older people in Western countries—a meta-analysis. Ageing Res. Rev. **12**(1), 339–53 (2013)
140. D.J. Watts, S.H. Strogatz, Collective dynamics of 'small-world'networks. Nature **393**(6684), 440 (1998)
141. J.K. Wynn, C. Sugar, W.P. Horan, R. Kern, M.F. Green, Mismatch negativity, social cognition, and functioning in schizophrenia patients. Biol. Psychiat. **67**(10), 940–7 (2010)

Chapter 8
Analysis of EEG in Medically Intractable Epilepsy

Ki-Young Jung

Abstract EEG is an electrophysiologic technique that directly measures neural activity from the brain with high temporal resolution. EEG has been used in various neurological disorders to monitor functional brain states such as impaired consciousness including brain death, epilepsy, sleep disorders and cerebrovascular diseases. Epilepsy is a neurological disorder characterized by producing epileptic seizure caused by abnormal excessive or synchronous neuronal activity in the brain. EEG is an essential method to make diagnosis, classification, and localization of seizure focus in epilepsy, and is actually most commonly used in epilepsy and seizure disorders. In this chapter, I'll focus on how EEG analysis can contribute to the clinical diagnosis and treatment of epilepsy, particularly to epilepsy surgery. As detailed theoretical and technological aspects of EEG analyses are introduced in other chapters, only clinical applicability and significance of these methods will be reviewed here.

8.1 Overview of Epilepsy and Epilepsy Surgery

Epilepsy is a chronic, neurologic disorder characterized by spontaneous, recurrent seizures. The prevalence of epilepsy is ranging from 2 to 10 per 1000 people, affecting approximately 70 million people of all ages worldwide [25]. About 20–40% of people with epilepsy are medically intractable or drug-resistant despite that a variety of antiepileptic drugs have been developed for the past two decades. Epilepsy surgery has been the gold standard therapeutic option for cases in which medications have failed [31]. About two third of patients who had surgical treatment exhibited good outcome [44]. The aim of resective epilepsy surgery is to achieve freedom of seizure by means of complete elimination of epileptogenic zone in the epileptic brain. Epileptogenic zone is a theoretical concept indicating brain area that is necessary and sufficient for initiating seizures and whose removal or disconnection is necessary for abolition of seizures [12]. We can only identify epileptogenic zone after achieving

K.-Y. Jung (✉)
Department of Neurology, Seoul National University, Seoul, Republic of Korea
e-mail: jungky@snu.ac.kr

C.-H. Im (ed.), *Computational EEG Analysis*, Biological and Medical Physics, Biomedical Engineering, https://doi.org/10.1007/978-981-13-0908-3_8

seizure-free by resection of brain region presumed to be epileptogenic zone. Since epileptogenic zone cannot be measured directly, its location should be inferred indirectly by defining other zones including irritative zone, epileptogenic lesion, and symptomatogenic zone, and functional deficit zone [35].

Presurgical evaluation for epilepsy include high-resolution magnetic resonance imaging (MRI) scan and video-EEG monitoring. Functional neuroimagings such as single photon emission tomography (SPECT) and positron emission tomography (PET) are frequently used as ancillary tools especially in case of no discernible lesion on brain MRI. *Seizure onset zone* (SOZ), or ictal onset zone, is the area of cortex from which seizures are generated, which can be measured by noninvasive and invasive EEG recordings. As epileptogenic zone involves SOZ, accurate localization of SOZ is imperative for successful seizure control. Scalp EEG is noninvasive electrophysiologic test which helps to identify SOZ. However, the accuracy of scalp EEG for localization or lateralization is reported less than 50% [18, 24, 26].

Invasive EEG recording using intracranial placement of electrodes such as depth electrode, subdural strip and grid is necessary when discordant findings among presurgical evaluations are present or epileptogenic zone cannot be determined by surface EEG recording. Intracranial EEG provides much higher spatial information, which is able to pick up potential changes occurring over only a few millimeters of cortex. However, apart from invasiveness, the spatial sampling is restricted by the number of electrodes, which may lead to fail to identify SOZ [37]. Despite the invasiveness, localization of SOZ by intracranial EEG monitoring is still regarded as a gold standard. Brain MRI technology is developing rapidly and has huge impact on the diagnosis of various neurological diseases such as stroke, epilepsy, and dementia. It can provide us with high spatial resolution with accurate location and precise information about structural alterations in the brain. However, because of the poor temporal resolution of MRI, it cannot tell much about information on rapidly time-varying processes in the brain such as epileptic discharges and neurocognitive processes.

Thanks to the recent advances in computer technologies, EEG in conjunction with neuroimaging technologies have allowed us to extend its clinical utility for the evaluation of patients with epilepsy [33]. Application of EEG source imaging (ESI) and high frequency oscillation (HFO) for the presurgical evaluation have had significant impact on the identification of epileptogenic zone and understanding epileptogenesis. In this chapter, clinical utilities of ESI and HFO in the presurgical evaluation of epilepsy will be discussed. I will review clinical trials or applications of these techniques with relatively large samples, and discuss how these methods contribute to the surgical treatment of epilepsy.

8.2 EEG Source Imaging

ESI is a model-based imaging technique that integrates temporal and spatial components of EEG to identify the sources generating electrical potentials recorded on the scalp [22]. ESI has been introduced in the late 1990s and anecdotally used in the presurgical evaluation in a few large epilepsy surgery centers. At that time, simple spherical head model with a standard MRI template and lower numbers of electrodes (usually less than 32 channels) were used, and thus it usually provided only a coarse estimation of source locations in the standard brain models. With advanced computing technology and power, high density electrodes (up to 256), and realistic head model constructed using patient's own MRI data can now be used for ESI, and thus the accuracy and feasibility of ESI have been significantly increased.

In spite of the advances in computational and integrative imaging, ESI has not become routine everywhere [27]. One of the important problems is that clinical utility of ESI was largely limited because of the complicated methods without standardization and additionally required workforce and time [30]. A survey conducted among 25 European epilepsy surgery centers showed that ESI was performed by 12 centers: exclusively with magnetoencephalography (MEG) in 3 centers, exclusively with EEG in 5 centers, and with both MEG and EEG in 4 centers. Furthermore, a total of 14 different combinations of inverse methods and volume conduction models were used: 7 for MEG and 13 for EEG [30]. Therefore, it is apparent that considerable gap between technological advancement and clinical utility exists in the field of ESI. At present, however, ESI seems to be a promising technique that can positively contribute to visual EEG analysis for the localization of epileptic spikes, and plays a role in epilepsy surgery evaluation [22]. ESI can be applied on either interictal epileptiform discharges (IEDs) or ictal discharges.

8.2.1 Interictal ESI

The IED is transient waves or complexes clearly distinguishable from background activity, and generally shows a pointed peak with a duration of 20–200 ms usually followed by a slow wave. The potential should be reflected in physically adjacent electrodes and perhaps in synaptically linked regions such as the contralateral hemisphere [14]. The IEDs are not accompanied by clinical or subclinical seizure. The IEDs are generated by the synchronous discharges of a group of neurons in a region referred to as the epileptic focus. IEDs are so highly correlated with spontaneous seizures that their presence is used to support the diagnosis of epilepsy and are considered as hallmark of epilepsy [41]. Thus, evaluation of the interictal EEG is an integral part of the presurgical evaluation.

At least 10–20 cm^2 of synchronous active cortex is usually necessary to produce a recognizable scalp potential. In other word, cortical areas that are active but small cannot be detected by the scalp electrodes due to the low conductivity of intervening tissues especially skull between cerebral cortex and scalp. Spikes arising from deep cortical areas cannot be detectable in scalp EEG until it propagates to more superficial areas. The site of IEDs represents irritative zone, and it can be extensive, usually much larger than epileptogenic zone [11]. Patients with focal IEDs included in surgical resection have good surgical outcomes. The presence of IEDs extending beyond the area of resection correlates with poor surgical outcome in patients with extrahippocampal epilepsy [3]. Thus, identification of interictal spike area is helpful for surgical treatment of epilepsy.

There are several studies reporting various levels of accuracies of interictal ESI. Inconsistent results may be partly attributed to the complicated steps for ESI and absence of standardized methods for ESI. The processes for ESI include generation of head model from brain MRI, calculation of inverse solution, and co-registration of electrode coordinates with brain MRI (refer to Fig. 1 in Brodbeck et al. [8]). There are various methodologies that can be possibly adopted for ESI, in terms of brain MRI (individual vs. standard template), head model (spherical, boundary element method, and finite element method), number of electrodes, electrode coordinates (individual vs. standard), and inverse methods (discrete vs. distributed). As numerous methods can be generated from the combination of the above-mentioned methods at each step of ESI, the diverse results are inevitable.

Furthermore, a variety of gold standards for estimating accuracy of ESI have been adopted, such as phantom simulation, comparison with spike location using intracranial electrodes, distance from surgical resection margin, and postoperative surgical outcome. For these reasons, care should be taken not to misinterpret the results of ESI. The best way to assess clinical impact of ESI on epilepsy surgery seems to evaluate the correlation of the results of ESI with prospective long-term (a minimum 2 years follow-up) postoperative surgical outcomes. There are only few studies referring to surgical outcomes.

Brodbeck and his colleagues prospectively evaluated clinical utility of ESI as a part of the pre-surgical work-up from 152 patients with a variety of partial epilepsies who underwent surgical treatment for intractable seizures [8]. Two third of patients had temporal lobe epilepsy and 77% had good surgical outcome, which was assessed at least 1 year after surgery. Standard clinical EEG using less than 32 electrodes was obtained basically, and a high-resolution EEG with 128 or 256 electrodes was also recorded in 55 patients. However, the method for coordination of electrode location (i.e., measured vs. standard template coordinate) was not described. A simple realistic head model using spherical model with anatomical constraints (SMAC) method was generated from individual MRI, and the linear distributed inverse algorithm known as local autoregressive average (LAURA) was used to estimate intracranial sources for IEDs. The rising phase of averaged IED from the most prevalent IEDs was subject to the source estimation.

Good surgical outcome (Engel class I and II) was considered to be the gold standard for correct localization of the epileptogenic focus. *Sensitivity* was defined as the percentage of patients with focus localization within the resected zone of all patients who were seizure-free (n = 117). *Specificity* was defined as the percentage of patients with localized focus outside the resected zone of all patients who had an Engel Class III or IV outcome after surgery (n = 19). Sensitivity and specificity were compared among various source estimation conditions, and among other neuroimaging methods. The sensitivity and specificity of ESI were highest when high resolution EEG and individual MRI were adopted, which showed 84.1 and 87.5%, respectively. When compared to other imaging modalities, these values were higher than those of structural MRI (76% sensitivity, 53% specificity), PET (69% sensitivity, 44% specificity) and ictal SPECT (58% sensitivity, 47% specificity). The accuracy of ESI was higher in case of temporal lobe epilepsy than extratemporal lobe epilepsy, although the difference did not reach statistical significance. The sensitivity of ESI in extratemporal lobe epilepsy was 75%, which was sharply contrasted to clinical localization of scalp EEG in which the diagnostic sensitivity of IEDs was reported to 37.1% in a large series of surgical cases of neocortical epilepsy [24].

The same group has extended their study to determine which procedure or combination of procedures is the most predictive of seizure-free outcome in patients undergoing epilepsy surgery in 190 patients with partial epilepsy patients [23]. The predictive value of non-invasive techniques including high-density electric source imaging (HD-ESI), MRI, PET, and SPECT in terms of sensitivity, specificity, predictive value and diagnostic odds ratio (OR) in relation to postoperative outcome which was evaluated at least 12 months after surgery (mean 26.6 ± 27 months). EEGs were obtained with 64, 128, or 256 electrodes in 85 patients. Averaged spikes were used and ESI was estimated by LAURA. Among 58 patients who underwent all tests, the adjusted prognostic OR was 13.1 for HD-ESI and 10.9 for MRI (both $p = 0.004$). 92.3% were seizure-free when combination of MRI and HD-ESI were positive, and none was seizure-free if both tests were negative.

Feng et al. [13] reported accuracy of ESI in forty-three patients with temporal lobe epilepsy (age ranged from 9 to 48 years old) [13]. ESI was estimated from EEG recording acquired with high density 256-channel. Forward model was a realistic atlas head model based on the finite difference method (FDM), and Low resolution electromagnetic tomography (LORETA) method was used for inverse estimation. Postoperative surgical outcome was used as a gold standard for epileptogenic zone (follow-up duration: minimum 7 months, median = 14 months, average = 13.6 months). 35 patients had good surgical outcome (Engel class I or II), and eight patients had Engel class III or IV. Sensitivity and specificity were measured with reference to surgical resection margin. At the sublobule level, the sensitivity and specificity of high density ESI were 91.4% and 75%, respectively, which showed the best performance compared to other noninvasive neuroimaging modalities (PET: 80% sensitivity and 50% specificity, MRI: 77.1% sensitivity and more than 60% specificity). The accuracy of high density ESI was comparable to that of the previous study [8].

It is remarkable that they investigated the relationship between the high density ESI source patterns and surgical outcome with the Kaplan–Meier survival analysis which showed that cases with *single source* had better prognosis than cases with *multiple sources* (88.9% vs. 42.9%, respectively). Furthermore, surgical prognosis was better when sources are present within the resection margin than when sources are beyond the resection margin (94.1% vs. 33.3%, respectively). However, they did not report how many patients had lesion on MRI, and how many patients had mesial or lateral temporal lobe epilepsy. It was not also reported how many patients were of children.

ESI can also be successfully applied on children with epilepsy. Russo et al. [36] retrospectively evaluated the diagnostic utility of ESI in 60 pediatric epilepsy patients who were younger than 18 years old [36]. EEG data were recorded with electrodes less than 32 channels (conventional 10–20 system plus 4–10 extra-channels over the suspected epileptogenic region). ESI was done with the realistic subject-dependent head models with a boundary element model (BEM) using a single rotating dipole for each spike or sharp wave. Accuracy of ESI was evaluated whether the single dipole was localized within surgical resection cavity after at least 1 year of follow-up. Epileptogenic focus in 65% of patients could be localized by single rotating dipole with low-resolution EEG, which was comparable to ictal SPECT (68%) and superior to PET (55.5%). Correct localization was significantly higher in case of temporal lobe epilepsy compared to extratemporal lobe epilepsy (84.6% vs. 48.0%). MRI-negative patients showed tendency to be more frequently localized within surgical resection cavity compared to MRI-positive cases (78.6% vs. 60.9%). Sensitivity and specificity of ESI in relation to 2-year surgical outcome were 60.6 and 50.0%, respectively. Although these figures are somewhat lower than those of previous reports, it is acceptable and reasonable considering lower number of electrodes adopted in this study, compared to high density (from 64 to 256 channels) EEG. Nonetheless, low-resolution electrode ESI was superior to PET or ictal SPECT images in the present study. Therefore, this study is encouraging that using only low-resolution EEG can be helpful in presurgical evaluation of intractable focal epilepsy at least with pediatric patients.

In summary, interictal ESI can be a relatively inexpensive, noninvasive modality for the evaluation of focal intractable epilepsy, and should be encouraged to use in routine presurgical evaluation in clinical practice. However, standardization of ESI processing steps and determination of IEDs should be provided for use in routine clinical practices. More strong data such as data from double-blinded randomized controlled study with large sample size are also needed.

8.2.2 Ictal ESI

Ictal EEG activity is characterized as a run of EEG rhythm that changes the EEG background and evolves over time in waveform frequency, amplitude, and morphology [28]. Ictal EEG is usually accompanied by clinical seizure. Rarely, no seizure can be associated with apparent ictal discharges, so-called subclinical EEG seizure, which occurs in cases when very small cortical areas are synchronized, which is only noted during electrocorticography rather than scalp EEG, and in cases of simple partial seizures that do not evolve complex partial or generalized seizure.

Ictal onset zones are areas of cortex where seizures are generated including areas of early propagation under certain circumstances, and can be identified by the source localization at ictal EEG onset time. Ictal onset zone can be either within the distribution of irritative zone or distinct from irritative zone. In other words, ictal onset zone is not necessarily the same as irritative zone. Nevertheless, the ictal ESI results are largely in agreement with interictal ESI results in temporal lobe epilepsy [34].

Since the aim of resective epilepsy surgery is to achieve freedom of seizure by complete elimination of epileptogenic zone, ictal EEG is considered to be more important than IEDs to determine the resection margin in surgical treatment of epilepsy. However, ictal ESI is largely limited by several reasons. For example, in a series of 100 presurgical candidates, ictal dipole source localization could be performed in only 31% of patients [5]. One of the reasons is that seizures with vigorous motor activities inevitably accompany many muscle and movement artifacts which lower signal-to-noise ratio hampering proper source localization. Ictal EEG activities often show bilateral or nonlateralized pattern, which also makes ESI difficult to correctly localize SOZ. It is not uncommon that patient do not have seizure during long-term EEG monitoring. Another important reason is that ictal rhythm can only be seen at the scalp when the cortical areas are sufficiently synchronized. Furthermore, seizures originating from medial side of cortex or deep sulci cannot be picked up by scalp electrodes until ictal activities propagate to lateral or superficial cortices. This means that ictal ESI identifies not only the ictal onset zone but also the cortex to which seizure discharges spread during an early ictal event [22].

Contrary to interictal ESI, there are relatively few clinical studies on the source localization of the ictal epileptiform activity. Source localization of rhythmic ictal scalp EEG activity has been reported on 42 patients with focal epilepsy, aged between 9 and 69 years (mean 35.9, median 38 years) by Danish group [4]. EEG was recorded using a standard 25 electrode (19 electrodes of the international 10–20 system with extra-electrodes in the inferior temporal chain). For the sake of clinical feasibility, standard electrode positions and template MRI (MNI template) were used for source localization in this study. The earliest rhythmic ictal activities that are stable in terms of frequency identified by spectral analysis and spatial distribution identified by voltage map were selected for the ictal ESI. The duration of these epochs ranged between 1 and 11 s (median 3 s). LAURA was used for the source localization. The predictive values were estimated based on the surgical outcome, as evaluated 1 year after the operation.

Ictal source localization at the sublobar level could be performed in 72% of the patients who had at least one epileptic seizure in the presurgical evaluation. The sensitivity of the ictal ESI was 69.7% and the specificity was 75.7% with regard to reference standard that was determined by the consensus conclusion of the multidisciplinary epilepsy surgery team. It has been reported that the seizure onset at the lobar level with reference to 2-year surgical outcome was correctly localized in 76–83% of temporal seizures and 47–65% of extratemporal seizures by visual inspection of scalp ictal EEG [43]. However, only 21–38% of scalp ictal EEG agreed with depth EEG at the sublobar level [38]. Therefore, it is remarkable that ictal ESI can significantly increase the accuracy of localization with higher precision than visual interpretation of scalp ictal EEG. Twenty patients underwent surgery and 16 patients (80%) became seizure-free. The PPV of the ictal source localization was 92%, and the NPV was 42.8%. Furthermore, the positive likelihood ratio for the concordant results (ictal source localization matching the reference standard) was 3.0 which is the nine times higher than the negative likelihood ratio of the discordant results (0.33, mismatch between ictal source localization and the reference standard).

In another study, ictal source localization using high-density scalp EEG data (128- or 256-electrode) was reported in 14 patients with focal epilepsy (eight male, mean age 26.3 years, range 10–45 years) [32]. Among them, 8 patients underwent epilepsy surgery and were followed up for at least 1 year after surgery. Dominant frequency of EEG rhythm was filtered with narrow band (±1 Hz) from epochs of early rhythmic ictal changes (mean 2 s, range 1.5–2.5 s), which was used for the source localization. The locally spherical model with anatomical constraints (LSMAC) for head model and LORETA for inverse method were used in this study. LORETA was performed at each time point of the filtered ictal EEG data and then averaged across the defined early ictal epoch. Six of eight patients who underwent epilepsy surgery were seizure free. Five of them showed the maximum current density of ictal ESI within the resected area, indicating ictal ESI correctly localized the SOZ in the resection area. On the other hand, two of the operated patients who had a less favorable outcome revealed that the maximal ictal ESI solution was not included in the resected area. Remarkably, the agreement between interictal and ictal ESI were concordant (9) or partially concordant (4) in 13 of 14 patients (93%).

Although there are still limited number of studies, ictal ESI provides more direct information about SOZ compared with interictal ESI (providing irritative zone), and it is feasible in clinical practice. Larger studies are needed to determine the clinical feasibility of ictal ESI as a valuable presurgical evaluation tool. Standardization and automation of processing steps, and multimodal approaches combining electrophysiology and neuroimages are strongly required for wider application of ictal ESI in clinical practice.

8.3 High Frequency Oscillation (HFOs)

The frequencies of conventional EEG recordings range from 0.5 to 70 Hz. However, high frequency EEG oscillatory activities (HFOs) above 70 Hz have been recognized during physiological and pathological conditions in intracranial recordings [9, 15]. HFOs refer to distinct types of brain activity occurring in a frequency band range between 80 and 500 Hz. Since HFO was discovered in epileptogenic hippocampus in patients with mesial temporal lobe epilepsy and in animal model of epilepsy using microelectrodes [6, 7], it has been considered to be a new biological marker for potential epileptogenecity although more convincing evidences are still needed.

Conventionally, IEDs have been considered as biological markers of epilepsy because almost all patients with epilepsy show IEDs at some time points. However, it is not uncommon that IEDs can be seen in healthy people or in some patients with neuropsychiatric disorders such as attention deficit hyperkinetic disorder, autism spectrum disorder, and developmental disorders. In addition, IEDs represent only irritative zones of cerebral cortex, neither epileptogenic zones nor SOZs. The identification of HFOs in epileptogenic tissues is one of the major discoveries in epilepsy research over the past two decades [21], because no clearly identified biological marker for epilepsy has been developed yet. HFOs are now increasingly used to identify and localize epileptogenic zones in presurgical evaluation of intractable epilepsy patients.

HFOs are subdivided into ripples ranging from 80 to 250 Hz and fast ripples >250 Hz [16]. Cellular and network mechanisms generating ripples and fast ripples remain unclear. Ripples reflect fast inhibitory postsynaptic potentials (IPSPs) on the soma of pyramidal cells. Ripples are observed in non-epileptic hippocampi and entorhinal cortices, and may play a role in memory consolidation by phase synchronization [20]. On the other hand, fast ripples are found only in the epileptogenic region and may reflect pathological hypersynchronous population spikes of bursting pyramidal cells.

8.3.1 What Are HFOs and How to Recognize HFOs

HFOs are defined as all physiologic and pathologic oscillatory EEG activities within a limited frequency band from 80 to 500 Hz range that clearly stand out from the baseline and persist for at least four oscillation cycles [49].

To record HFOs, there should be special considerations in terms of both hardware and software. In the beginning of HFOs discovery, HFOs were detected using microelectrodes (40 μm wires) implanted in the hippocampus and entorhinal cortex. Then, it has been also reported that HFOs can be picked up by macro-intracranial electrodes such as subdural and depth electrodes, which are clinically used to localize SOZ in presurgical evaluation. HFOs can be recorded even with scalp EEG. Therefore, it seems that any clinically used electrodes can now record HFOs.

The sampling rate of EEG amplifiers should be at least 4 times higher than the upper frequency of interest. For identifying ripples (80–250 Hz), proper sampling rate of amplifiers should be at least 1000 Hz, and a sampling rate of 2000 Hz or above should be used to record HFOs over 250 Hz [48]. Software for reviewing data should have a capability of a high-order finite impulse response (FIR) filter, and of extended time scale up to 0.6 s/page, and of amplitude scale changeable to 1 uV/mm [48, 49].

8.3.2 Interictal HFOs

The first recording of HFOs during interictal period in human mesial temporal epilepsy with or without hippocampal sclerosis was performed by the University of California at Los Angeles group [39, 40]. Depth electrodes were inserted in both hippocampi and entorhinal cortices (i.e., epileptic side and healthy side). They found two distinct oscillatory groups composed of the lower-frequency ripple group having a mean frequency of 96 Hz and the higher-frequency fast ripple group having a mean frequency of 262 Hz. The ratio of the number of fast ripples to ripples was significantly higher in the sites ipsilateral to seizure onset with hippocampal atrophy

Table 8.1 Comparison of sensitivity, specificity and accuracy with regard to detecting seizure onset zone

	Intracranial[a]			Scalp[b]		
	Spike (%)	Ripple (80–250 Hz) (%)	Fast ripple (>250 Hz) (%)	Spike (%)	Gamma (40–80 Hz) (%)	Ripple (>80 Hz) (%)
Sensitivity	91	91	66	100	82	48
Specificity	30	42	80	30	68	89
Accuracy	44	54	76	43	70	81

[a]Data from [1]
[b]Data from [2]

compared with sites contralateral to both seizure onset and hippocampal atrophy. These findings indicate the strong association between fast ripple and epileptogenic zone.

Interictal HFOs were observed in neocortical epilepsy with and without lesion in addition to limbic epilepsy which was firstly reported by Worrell et al. [45]. They recorded EEG activity using clinical subdural and depth electrodes. They found that HFOs (60–100 Hz) confined in the contacts of SOZ in all six patients. Thus, HFOs is identified not only in mesial temporal lobe epilepsy but also commonly observed in the focal neocortical epilepsy.

It is worth comparing spatiotemporal relationship between interictal HFOs and IEDs. It has been repeatedly reported that the interictal HFOs were consistent with the SOZ. HFOs are more frequently identified in the SOZ than outside SOZ [1, 2, 10]. However, HFOs are not limited to the SOZ but usually extended beyond seizure generating area. HFOs occurred to a large extent independently of spikes [19]. Although HFOs are found in spiking as well as nonspiking channels, HFOs generating zones have usually smaller spatial extent than the irritative zones [48]. HFOs often occur with close association with IEDs in terms of temporal relationship. 81% of HFOs occurs with IEDs, mostly superimposed over epileptic spikes [42], 19% occurred completely independently of spikes in timing and localization. In another study, 63% of ripple co-occurred with IEDs [2].

Andrade-Valença et al. evaluated specificity, sensitivity and accuracy of interictal HFOs and IEDs to determine the SOZ from 17 patients with intractable focal seizures with normal MRIs during 5–10 min of sleep recording of intracranial EEGs. The accuracy of HFOs (both ripple and fast ripple) for localizing SOZ was higher compared with that of IEDs [1]. Interictal HFOs can be identified in scalp EEG, and the accuracy of gamma (40–80 Hz) and ripple oscillations (>80 Hz) are higher compared with IEDs [2]. Thus, it seems that both intracranial and scalp recorded fast oscillations are less sensitive but much more specific and accurate than IEDs to delineate the SOZ (Table 8.1).

Taken together, interictal HFOs are more specific and accurate than IEDs to delineate the SOZ. HFOs can be considered as a reliable marker for SOZ, better than IEDs.

8.3.3 HFOs at Seizure Onset (Ictal HFOs)

HFOs occur during preictal and seizure onset periods. It is noted that high-frequency activity (60–100 Hz) increases significantly in the 20 min prior to neocortical seizure onset in 62% of all seizures [46].

HFOs were identified at the time of seizure onset in 7 of the 10 patients with focal epilepsy [20]. HFOs were present in very limited distributions with high frequency events (100–200 Hz) confined to 1–3 contacts and with very high frequency events (250–500 Hz) confined to 1–2 contacts of the same depth electrode. The distribution of HFOs correlates well with the SOZ determined by the comprehensive presurgical evaluation [20].

Modur et al. demonstrated HFOs (>70 Hz) at seizure onset (ictal HFOs) in six patients with neocortical epilepsy. Ictal HFOs had a widespread activity at seizure onset but evolved subsequently in a restricted manner (HFOs+). HFOs+ occurred earlier and with a smaller distribution than the seizure onset defined by the conventional frequency activity (<40 Hz). Furthermore, five out of six patients (83%) showed favorable postoperative outcome over a mean follow-up of 27 months, which included mainly the HFOs+ channels [29].

HFOs show a remarkable reliability as it appears in the same channels during interictal, preictal and ictal periods in depth electrode recordings. On the other hand, epileptic spikes are more widespread and occur on different channels in ictal and interictal periods [47]. Therefore, HFOs are most likely linked to epileptogenesis.

8.3.4 HFOs and Surgical Outcome

The clinical utility of HFOs has been exemplified by better surgical outcome when resecting cortical areas generating HFOs compared with resecting conventional SOZ. Cho et al. evaluated the occurrence rate and extent between HFO distribution and the SOZ in 15 patients with neocortical epilepsy. They found that resection of high-rate HFO regions was significantly associated with favorable outcome. Interestingly, the extent/ratio of SOZ or spiking region resection did not differ between seizure-free and non-seizure-free groups [10].

A meta-analysis which included 11 studies investigated surgical outcomes based on resection ratio of HFOs, that is the ratio between the number of channels on which HFOs were detected and, among these, the number of channels that were inside the resected area [17]. Postoperative good surgical outcome is related to a higher resection ratio of HFOs compared with poor surgical outcome. The effect sizes are significant but small. Therefore, randomized controlled clinical trials are required to make convincing evidences in clinical practice.

8.3.5 Illustrated Case

This is a case of a 36-year-old man with epilepsy who showed the usefulness of HFO in epilepsy surgery. He was right-handed and had no febrile convulsion episode, and his seizures started at age of 22 years. The patient had no previous medical history other than epilepsy. The patient had three types of seizure semiology: (1) rotation rightward without impaired awareness and slightly impaired responsiveness, (2) automotor or hypermotor seizure and (3) secondary generalized tonic-clonic seizure. He had seizures two times a week while taking multiple AEDs including carbamazepine 1200 mg, valproate 1200 mg, topiramate 200 mg and levetiracetam 1000 mg daily divided in two doses. Brain MRI, PET and interictal SPECT revealed no definite abnormalities.

In scalp video-EEG monitoring, interictal spikes were observed in right frontal and midline (F4 and Fz maximum) areas (Fig. 8.1). Several episodes of automotor seizures mainly involving right leg were observed and ictal EEG showed that (1) both frontal and temporal onset rhythmic activities, or (2) left temporal onset rhythmic

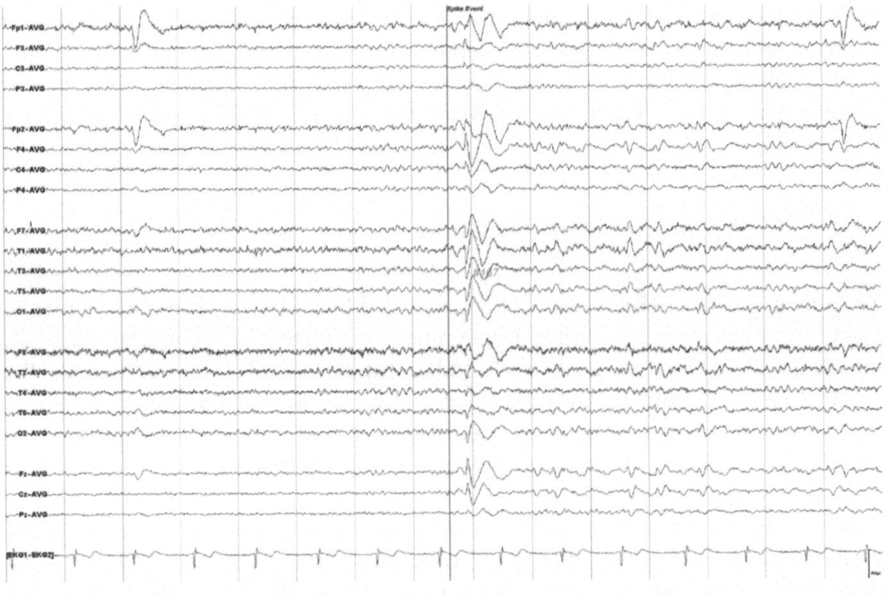

Fig. 8.1 Scalp EEG shows interictal spikes over frontal region. F4 and Fz electrodes shows maximum negativity against the common average reference

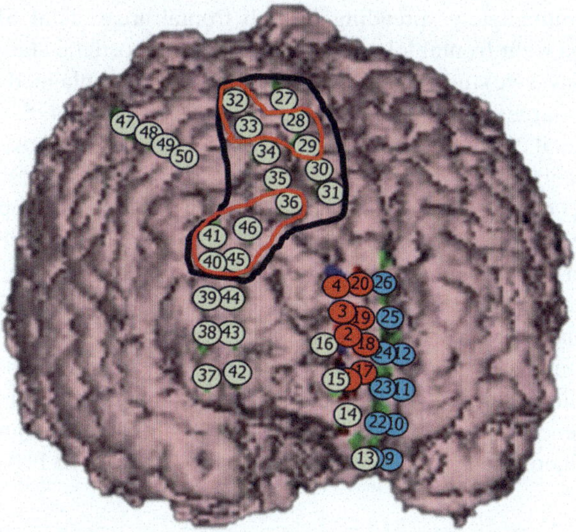

Fig. 8.2 Inserted electrodes. Red circles show the irritative zones and a black circle indicates the recommended resection margin

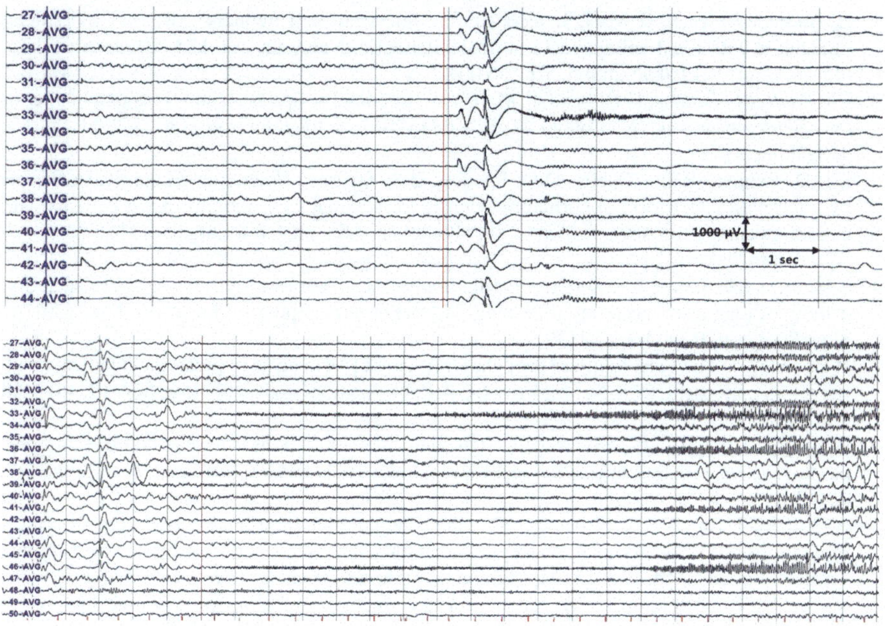

Fig. 8.3 Invasive EEG: (upper) Interictal EEG, (lower) Ictal EEG. Red vertical line shows the onset of seizure. The frequency of HFO in channel 33 was around 70 Hz after the seizure onset, and decreased to around 30 Hz after 5 s

theta activity, immediately spreading to both frontal areas. Ictal SPECT showed hypoperfusion in right frontal lobe, possibly indicating postictal status.

Subsequent invasive study covering right frontal areas was planned (Fig. 8.1). Two depth electrodes targeting right medial frontal area, (channel 1–4, 5–8) two grids on right medial frontal cortex, (channel 9–12, 21–26) and four grids covering most area of right frontal cortex (channel 13–16, 27–36, 37–46, 47–50) were applied.

Invasive EEG revealed interictal spikes and HFOs in posterior upper frontal lobe (channel 28, 29, 32, 33, 36, 40, 45, 46; Fig. 8.2). Especially, ictal EEG started as a form of HFO in channel 33 (precentral gyrus in upper frontal lobe) and spread to neighboring channels (Fig. 8.3). The corticectomy including channel 33 and irritative zones with medial frontal cortex was performed. During 20 months after the surgery, the patient remained seizure free maintaining only the carbamazepine 800 mg daily in out-patient clinic. Cortical functioning was also significantly improved when the general intelligence was evaluated using Full Scale Intelligence Quotient with the resultant score increasing from preoperative 88 to postoperative 110.

References

1. L. Andrade-Valença, F. Mari, J. Jacobs et al., Interictal high frequency oscillations (HFOs) in patients with focal epilepsy and normal MRI. Clin. Neurophysiol. **123**, 100–105 (2012). https://doi.org/10.1016/j.clinph.2011.06.004
2. L.P. Andrade-Valenca, F. Dubeau, F. Mari et al., Interictal scalp fast oscillations as a marker of the seizure onset zone. Neurology **77**, 524–531 (2011). https://doi.org/10.1212/WNL.0b013e318228bee2
3. R.E.D. Bautista, M.A. Cobbs, D.D. Spencer, S.S. Spencer, Prediction of surgical outcome by interictal epileptiform abnormalities during intracranial EEG monitoring in patients with extrahippocampal seizures. Epilepsia **40**, 880–890 (1999). https://doi.org/10.1111/j.1528-1157.1999.tb00794.x
4. S. Beniczky, G. Lantz, I. Rosenzweig et al., Source localization of rhythmic ictal EEG activity: a study of diagnostic accuracy following STARD criteria. Epilepsia **54**, 1743–1752 (2013). https://doi.org/10.1111/epi.12339
5. P. Boon, M. D'Havé, B. Vanrumste et al., Ictal source localization in presurgical patients with refractory epilepsy. J. Clin. Neurophysiol. Off. Publ. Am. Electroencephalogr. Soc. **19**, 461–468 (2002)
6. A. Bragin, J. Engel, C.L. Wilson et al., High-frequency oscillations in human brain. Hippocampus **9**, 137–142 (1999a). https://doi.org/10.1002/(sici)1098-1063(1999)9:2<137::aid-hipo5>3.0.co;2-0
7. A. Bragin, J. Engel, C.L. Wilson et al., Hippocampal and entorhinal cortex high-frequency oscillations (100–500 Hz) in human epileptic brain and in kainic acid-treated rats with chronic seizures. Epilepsia **40**, 127–137 (1999b). https://doi.org/10.1111/j.1528-1157.1999.tb02065.x
8. V. Brodbeck, L. Spinelli, A.M. Lascano et al., Electroencephalographic source imaging: a prospective study of 152 operated epileptic patients. Brain **134**, 2887–2897 (2011). https://doi.org/10.1093/brain/awr243
9. G. Buzsaki, Z. Horvath, R. Urioste et al., High-frequency network oscillation in the hippocampus. Science **256**, 1025–1027 (1992)
10. J.R. Cho, D.L. Koo, E.Y. Joo et al., Resection of individually identified high-rate high-frequency oscillations region is associated with favorable outcome in neocortical epilepsy. Epilepsia **55**, 1872–1883 (2014). https://doi.org/10.1111/epi.12808

11. J.S. Ebersole, D.R. Nordli, A.M. Husain, Current practice of clinical electroencephalography (2014)
12. J. Engel, *Seizures and Epilepsy*, 2nd edn. (Oxford University Press, New York, 2013)
13. R. Feng, J. Hu, L. Pan et al., Application of 256-channel dense array electroencephalographic source imaging in presurgical workup of temporal lobe epilepsy. Clin. Neurophysiol. **127**, 108–116 (2016). https://doi.org/10.1016/j.clinph.2015.03.009
14. R.S. Fisher, H.E. Scharfman, M. deCurtis, How can we identify ictal and interictal abnormal activity? Adv. Exp. Med. Biol. **813**, 3–23 (2014). https://doi.org/10.1007/978-94-017-8914-1_1
15. R.S. Fisher, W.R. Webber, R.P. Lesser et al., High-frequency EEG activity at the start of seizures. J. Clin. Neurophysiol. Off. Publ. Am. Electroencephalogr. Soc. **9**, 441–448 (1992)
16. B. Frauscher, F. Bartolomei, K. Kobayashi et al., High-frequency oscillations: the state of clinical research. Epilepsia **58**, 1316–1329 (2017). https://doi.org/10.1111/epi.13829
17. Y. Höller, R. Kutil, L. Klaffenböck et al., High-frequency oscillations in epilepsy and surgical outcome. A meta-analysis. Front. Hum. Neurosci. **9** (2015). https://doi.org/10.3389/fnhum.2015.00574
18. K.-S. Hong, S.K. Lee, J.-Y. Kim et al., Pre-surgical evaluation and surgical outcome of 41 patients with non-lesional neocortical epilepsy. Seizure **11**, 184–192 (2002). https://doi.org/10.1053/seiz.2001.0616
19. J. Jacobs, P. LeVan, R. Chander et al., Interictal high-frequency oscillations (80–500 Hz) are an indicator of seizure onset areas independent of spikes in the human epileptic brain. Epilepsia **49**, 1893–1907 (2008). https://doi.org/10.1111/j.1528-1167.2008.01656.x
20. J.D. Jirsch, E. Urrestarazu, P. LeVan et al., High-frequency oscillations during human focal seizures. Brain **129**, 1593–1608 (2006). https://doi.org/10.1093/brain/awl085
21. P. Jiruska, C. Alvarado-Rojas, C.A. Schevon et al., Update on the mechanisms and roles of high-frequency oscillations in seizures and epileptic disorders. Epilepsia **58**, 1330–1339 (2017). https://doi.org/10.1111/epi.13830
22. K. Kaiboriboon, H.O. Lüders, M. Hamaneh et al., EEG source imaging in epilepsy—practicalities and pitfalls. Nat. Rev. Neurol. **8**, 498–507 (2012). https://doi.org/10.1038/nrneurol.2012.150
23. A.M. Lascano, T. Perneger, S. Vulliemoz et al., Yield of MRI, high-density electric source imaging (HD-ESI), SPECT and PET in epilepsy surgery candidates. Clin. Neurophysiol. **127**, 150–155 (2016). https://doi.org/10.1016/j.clinph.2015.03.025
24. S.K. Lee, S.Y. Lee, K.-K. Kim et al., Surgical outcome and prognostic factors of cryptogenic neocortical epilepsy. Ann. Neurol. **58**, 525–532 (2005). https://doi.org/10.1002/ana.20569
25. S.-Y. Lee, S.-E. Chung, D.W. Kim et al., Estimating the prevalence of treated epilepsy using administrative health data and its validity: ESSENCE study. J. Clin. Neurol. **12**, 434–440 (2016). https://doi.org/10.3988/jcn.2016.12.4.434
26. S.-Y. Lee, J.-Y. Lee, D.W. Kim et al., Factors related to successful antiepileptic drug withdrawal after anterior temporal lobectomy for medial temporal lobe epilepsy. Seizure **17**, 11–18 (2008). https://doi.org/10.1016/j.seizure.2007.05.014
27. F.S.S. Leijten, G. Huiskamp, Interictal electromagnetic source imaging in focal epilepsy: practices, results and recommendations. Curr. Opin. Neurol. **21**, 437–445 (2008). https://doi.org/10.1097/WCO.0b013e3283081e23
28. Mayo Clinic College of Medicine, Rochester, Minnesota, St. E. Louis, L. Frey et al., Electroencephalography (EEG): an introductory text and atlas of normal and abnormal findings in adults, children, and infants. American Epilepsy Society (2016)
29. P. Modur, S. Miocinovic, Interictal high-frequency oscillations (HFOs) as predictors of high frequency and conventional seizure onset zones. Epileptic. Disord. **17**, 413–424 (2015). https://doi.org/10.1684/epd.2015.0774
30. B.E. Mouthaan, M. Rados, P. Barsi et al., Current use of imaging and electromagnetic source localization procedures in epilepsy surgery centers across Europe. Epilepsia **57**, 770–776 (2016). https://doi.org/10.1111/epi.13347
31. D.R. Nair, Management of drug-resistant epilepsy. Contin. Lifelong Learn Neurol. **22**, 157–172 (2016)

32. P. Nemtsas, G. Birot, F. Pittau et al., Source localization of ictal epileptic activity based on high-density scalp EEG data. Epilepsia **58**, 1027–1036 (2017). https://doi.org/10.1111/epi.13749
33. A.C. Patel, R.C. Thornton, T.N. Mitchell, A.W. Michell, Advances in EEG: home video telemetry, high frequency oscillations and electrical source imaging. J. Neurol. **263**, 2139–2144 (2016). https://doi.org/10.1007/s00415-016-8159-3
34. C. Plummer, A.S. Harvey, M. Cook, EEG source localization in focal epilepsy: where are we now? Epilepsia **49**, 201–218 (2008). https://doi.org/10.1111/j.1528-1167.2007.01381.x
35. F. Rosenow, H. Lüders, Presurgical evaluation of epilepsy. Brain **124**, 1683–1700 (2001). https://doi.org/10.1093/brain/124.9.1683
36. A. Russo, P. Jayakar, M. Lallas et al., The diagnostic utility of 3D electroencephalography source imaging in pediatric epilepsy surgery. Epilepsia **57**, 24–31 (2016). https://doi.org/10.1111/epi.13228
37. S.D. Shorvon, E. Perucca, J. Engel (eds.) Invasive EEG in presurgical evaluation of epilepsy, in *The Treatment of Epilepsy*, 3rd ed. (Wiley-Blackwell, Chichester, UK; Hoboken, NJ, 2009), pp. 767–798
38. S.S. Spencer, P.D. Williamson, S.L. Bridgers et al., Reliability and accuracy of localization by scalp ictal EEG. Neurology **35**, 1567 (1985). https://doi.org/10.1212/WNL.35.11.1567
39. R.J. Staba, C.L. Wilson, A. Bragin et al., Quantitative analysis of high-frequency oscillations (80–500 Hz) recorded in human epileptic hippocampus and entorhinal cortex. J. Neurophysiol. **88**, 1743–1752 (2002)
40. R.J. Staba, C.L. Wilson, A. Bragin et al., High-frequency oscillations recorded in human medial temporal lobe during sleep. Ann. Neurol. **56**, 108–115 (2004). https://doi.org/10.1002/ana.20164
41. K.J. Staley, F.E. Dudek, Interictal spikes and epileptogenesis. Epilepsy. Curr. **6**, 199–202 (2006). https://doi.org/10.1111/j.1535-7511.2006.00145.x
42. E. Urrestarazu, R. Chander, F. Dubeau, J. Gotman, Interictal high-frequency oscillations (100–500 Hz) in the intracerebral EEG of epileptic patients. Brain **130**, 2354–2366 (2007). https://doi.org/10.1093/brain/awm149
43. T.S. Walczak, R.A. Radtke, D.V. Lewis, Accuracy and interobserver reliability of scalp ictal EEG. Neurology **42**, 2279 (1992). https://doi.org/10.1212/WNL.42.12.2279
44. S. West, S.J. Nolan, J. Cotton et al., Surgery for epilepsy, in *Cochrane Database of Systematic Reviews* (Wiley, New York, 2015)
45. G.A. Worrell, K. Jerbi, K. Kobayashi et al., Recording and analysis techniques for high-frequency oscillations. Prog. Neurobiol. **98**, 265–278 (2012). https://doi.org/10.1016/j.pneurobio.2012.02.006
46. G.A. Worrell, L. Parish, S.D. Cranstoun et al., High-frequency oscillations and seizure generation in neocortical epilepsy. Brain **127**, 1496–1506 (2004). https://doi.org/10.1093/brain/awh149
47. M. Zijlmans, J. Jacobs, Y.U. Kahn et al., Ictal and interictal high frequency oscillations in patients with focal epilepsy. Clin. Neurophysiol. **122**, 664–671 (2011). https://doi.org/10.1016/j.clinph.2010.09.021
48. M. Zijlmans, P. Jiruska, R. Zelmann et al., High-frequency oscillations as a new biomarker in epilepsy. Ann. Neurol. **71**, 169–178 (2012). https://doi.org/10.1002/ana.22548
49. M. Zijlmans, G.A. Worrell, M. Dümpelmann et al., How to record high-frequency oscillations in epilepsy: a practical guideline. Epilepsia **58**, 1305–1315 (2017). https://doi.org/10.1111/epi.13814

Chapter 9
Computational EEG Analysis for Brain-Computer Interfaces

Garett D. Johnson and Dean J. Krusienski

Abstract EEG activity can be actively or passively modulated in a way to provide commands to external devices. The feedback provided by interacting with the EEG-controlled device creates a closed-loop system with the user in the loop. Such a system is known as a Brain-Computer Interface (BCI). The selection of an analysis approach for BCIs should be guided by the nature of the signals in consideration. This chapter presents the most fundamental and widely-used EEG analysis techniques organized by the type of control signal.

9.1 Brain-Computer Interfaces

9.1.1 Introduction

A Brain-Computer Interface (BCI) uses brain responses to deliberately-designed sensory stimuli or spontaneous mental activity to provide commands to external devices. A block diagram of a typical BCI is shown in Fig. 9.1. The digitized signals are commonly preprocessed, which includes preemptive elimination of known interference (i.e., artifacts) or irrelevant information, and/or the enhancement of spatial, spectral, or temporal characteristics of the signal that are particularly relevant to the application. The preprocessed signals are passed to the feature extraction stage, which can represent a variety of techniques for effectively isolating the relevant information in the signals for BCI control. Commonly, more than one feature are extracted from the signals and the resulting set of features for a given observation interval is processed as a feature vector. This feature vector is then passed to the classifier (or regressor),

G. D. Johnson · D. J. Krusienski (✉)
Department of Electrical and Computer Engineering, Old Dominion University, Norfolk, VA, USA
e-mail: dkrusien@odu.edu

C.-H. Im (ed.), *Computational EEG Analysis*, Biological and Medical Physics, Biomedical Engineering, https://doi.org/10.1007/978-981-13-0908-3_9

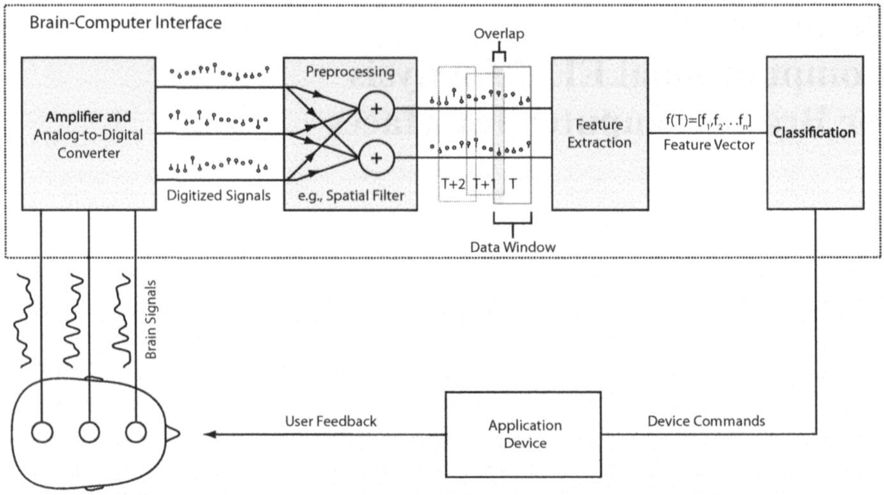

Fig. 9.1 Block diagram of a Brain-Computer Interface (BCI)

where it is converted into device commands and feedback to the user. It is important to note that in certain designs (e.g., artificial neural networks), a single transformation is used to convert the digital signals directly into device commands and there is no clear distinction between the feature extraction and classification stages.

Accurate and robust feature extraction can greatly simplify the subsequent classification, and produces more accurate and reliable actions and more natural feedback to the user. On the other hand, it is possible to compensate for somewhat poor or non-specific feature extraction by using a more complex classification algorithm to produce equally effective results. It is important to recognize that feature extraction and translation go hand-in-hand and that practical BCI systems balance the emphasis between these two stages to ensure that they work together effectively.

BCI systems for disabled users aim to facilitate communication and/or environmental interaction capabilities which have been lost or impaired by injury or disease. The most straightforward and flexible approach to satisfying the immediate needs of this disabled user population is to enable the user to control a personal computer. In this way, the user can control standard or custom software applications such as Internet browsers, word processors, and email; as well as interface with limitless external devices such as appliances, robotic arms, climate controllers, etc. Accordingly, most BCI outputs aim to mimic the two most ubiquitous computer input devices: the continuous dimensional control of a mouse and the discrete selection control of a keyboard. Control of such low degree-of-freedom devices is typically amenable to localized and low-density recordings (i.e., using few electrodes).

For a BCI keyboard-type output, the user is presented (visually, aurally, or tactilely) with a variety of selectable options, each representing a character, word, function, or an even more complex device command or series of commands. These options are selected using transient changes in the user's brain activity, which correspond to the task. Selection of one or more of these options in sequence allows the user to type messages, navigate menus, engage preprogrammed actions of an external device, etc. Continuous mouse-like control is achieved by translating the coordinated modulation of the user's brain activity into the intended directional control commands. This allows the user to navigate a cursor to icons on a computer screen, freehand draw, and can be directly extended to achieve continuous control of a robotic arm, a wheelchair, or other devices that require continuous dimensional control.

Although achieving reliable 1-, 2-, or 3- degree of freedom control is useful and sufficient for most assistive applications, a more ambitious objective is to design a BCI that attempts to replicate natural, high degree-of-freedom function in a manner that is more transparent to the user. For example, for communication, language cortex signals during imagined speech would be decoded by the BCI and replicated using a speech synthesizer in real-time. Similarly, for motor control, motor cortex signals during imagined limb movements would be decoded by the BCI and replicated using a prosthetic limb, orthosis, or even neuromuscular stimulation of the impaired limb in real-time.

Approaching transparent replication of natural function via a BCI proves to be very challenging for a variety of reasons. For instance, given the distributed complexities of motor and language processes in the brain, it is difficult to capture all of the subtle nuances needed to reliably reproduce completely natural function from limited recording sites (even on the order of thousands of single neuron or local field potential recordings). Related to this point, it is expected that only invasive recordings can provide the appropriate signals and resolution required to achieve this type of high-level intrinsic control. Additionally, other difficult issues such as the role of sensory feedback and proprioception in the replication of natural function must be considered. Consequently, simplified approaches that require fewer recording sites and less sophistication, such as limited vocabulary speech and limited degree-of-freedom motor commands, can serve as more practical alternatives as the technology continues to advance.

Any of a variety of brain signals can be translated by a BCI to achieve a particular device output. For example, brain activity from relevant language areas recorded while a user imagines vocalizing a word can be used to control a speech synthesizer. This same brain activity could also be used to control a hand orthosis, where the presence of a particular imagined word would close the orthosis and another imagined word would open it, for instance. Another example is a BCI that records from a single neuron (not necessarily from motor cortex) that has been conditioned to adjust the spike firing rate when the user wants to open/close the orthosis. Likewise, a signal generated over the sensorimotor cortex during imagined hand movement could be decoded by the BCI and used to open/close the hand orthosis corresponding to the imagery. This same sensorimotor signal could be used by a BCI in a communication application to select sequentially highlighted letters in a visual keyboard at the moment when the imagined movement is detected. The resulting typed message could also be synthesized as speech, thus achieving the same effective output as the first example using a completely unrelated signal and interface.

9.1.2 *Types of BCIs*

BCIs can generally be grouped into three categories: Reactive, Active, and Passive; based on how the brain response is elicited [34].

9.1.2.1 Reactive

The use of sensory stimuli to induce predictable changes in the EEG creates a "reactive" control paradigm wherein the EEG control signal and its detection are dependent upon the parameters of the sensory stimulus, which is typically visual, auditory, or tactile [2]. The P300 oddball paradigm and its extension, the P300 speller, are examples of reactive paradigms [10]. Such paradigms are commonly designed so that the EEG response differs depending on the characteristics of temporally and/or spatially multiplexed sensory stimuli, which individually represent various selection options for the control interface. In general, reactive paradigms are inherently synchronous systems, where timing of the user's intended device commands is dependent on the timing of the presented stimuli. Synchronous systems do not allow for completely spontaneous communication or control.

9.1.2.2 Active

In contrast to reactive paradigms, volitional mental imagery or cognitive state switching creates an "active" control paradigm wherein the changes in the EEG are due to endogenous cognitive actions and not external stimuli. An example of an active control paradigm is sensorimotor rhythm cursor control where the user imagines left/right hand movements to move a computer cursor to the left/right, respectively [32, 33]. Active paradigms are better suited for achieving asynchronous communication or control, although synchronous designs with cued mental imagery are also used.

9.1.2.3 Passive

The concept of passive BCIs where the user is not volitionally modulating the EEG for direct device control will not be presented in this chapter. However, most passive BCI scenarios and analyses are closely related to active/reactive paradigms and the techniques discussed in this chapter are directly applicable [35].

9.1.3 Types of EEG Responses for BCI Control

For achieving different modes of control, EEG responses can be further categorized into transient and continuous responses.

9.1.3.1 Transient

Transient responses are momentary changes in the EEG that can be associated with a transition of a sensory stimulus (reactive), or a transition in mental imagery or cognitive state such as a single hand grasp (active). Such transient responses are well-suited for making discrete selections in applications such as BCI typing.

9.1.3.2 Continuous

Continuous responses represent ongoing EEG activity that differs from resting-state activity. In reactive paradigms, this is associated with repetitive sensory stimuli that likewise produce repetitive "steady-state" patterns in the EEG. In active paradigms, continuous mental imagery or cognitive state entrainment produce sustained changes in the EEG that, in some cases, can also produce graded modulation with training to allow more refined control. Such continuous responses are well-suited for continuous dimensional-control applications such as controlling a cursor or a motorized wheelchair.

9.1.4 Organization of the Chapter

The selection of a processing approach must be guided by the nature of the signals in consideration. Most BCIs can be classified within the taxonomy of active/reactive and transient/continuous (see Table 9.1). Within this taxonomy, the conventional associated EEG analysis techniques are largely unique and distinctive. Thus, this chapter presents the most fundamental and widely-used EEG analysis techniques organized by this taxonomy.

9.2 Active Paradigms

The EEG responses associated with common active paradigms generally form distinct spatial patterns within a specific frequency range(s). Thus, techniques that detect the spatial patterns in frequency bands associated with mental imagery (i.e., μ and β) are commonly employed for this detection problem. Two fundamental approaches are

Table 9.1 Taxonomy of EEG responses

Response type	Trigger	Basic methods	Example
Active transient	Spontaneous mental imagery/state	Spatio-spectral analysis, common spatial patterns (CSP)	Single imagined hand grasp
Active continuous	Spontaneous mental imagery/state	Spatio-spectral analysis, CSP	Sustained motor imagery
Reactive transient	Transient sensory stimulus	Spatio-temporal discriminant analysis	P300 evoked potential
Reactive continuous	Repetitive sensory stimulus	Spatio-spectral analysis, canonical correlation analysis (CCA)	Steady-state visual evoked potential (SSVEP)

outlined. The first approach is an intuitive combination of data-independent spatial filtering and traditional spectral analysis, followed by a classification or regression model for producing the output command. The second approach, known as common spatial patterns, generates a data-dependent spatial filter that optimizes discrimination.

9.2.1 Traditional Spectral Analysis

9.2.1.1 Preprocessing

Assuming that the control signal is spatially-localized such as motor imagery (MI), it is prudent to employ a spatial filter such as a Large Laplacian over the relevant area(s) of the motor cortex to increase the SNR [22]. The Large Laplacian filter and associated weights are computed based on distance from the center electrode as follows:

$$V_i^{LAP} = V_i^{ER} - \sum_{j \varepsilon S_i} g_{ij} V_j^{ER} \tag{9.1}$$

where

$$g_{ij} = \frac{1/d_{ij}}{\sum_{j \varepsilon S_i} 1/d_{ij}} \tag{9.2}$$

refers to the ear-referenced voltage, S_i is the set of electrodes surrounding the i-th electrode, and d_{ij} is the distance between electrodes i and j (where j is a member of S_i). The Large Laplacian filter essentially acts as a data-independent beamformer by subtracting the average of surrounding electrodes from a central electrode of interest.

Referring to Fig. 9.2, the central electrode C3 is directly of the right-hand area of the sensorimotor cortex. Note how the spatial filter distinctly enhances the μ rhythm in both the time and frequency domains for improved detection and tracking.

Because MI is characterized in the μ (8–12 Hz) and β (18–24 Hz) frequency bands, it is common to bandpass filter between 0.1–5 Hz for the highpass cutoff and between 15–30 Hz for the lowpass cutoff, depending on the frequency band(s) used. Since the event is generally not time-locked to a stimulus, the EEG is evaluated in a continuous fashion over sliding data windows. Commonly sliding detection windows for BCI are in the range of 0.5–1 s with update rates of 0.5 s or less.

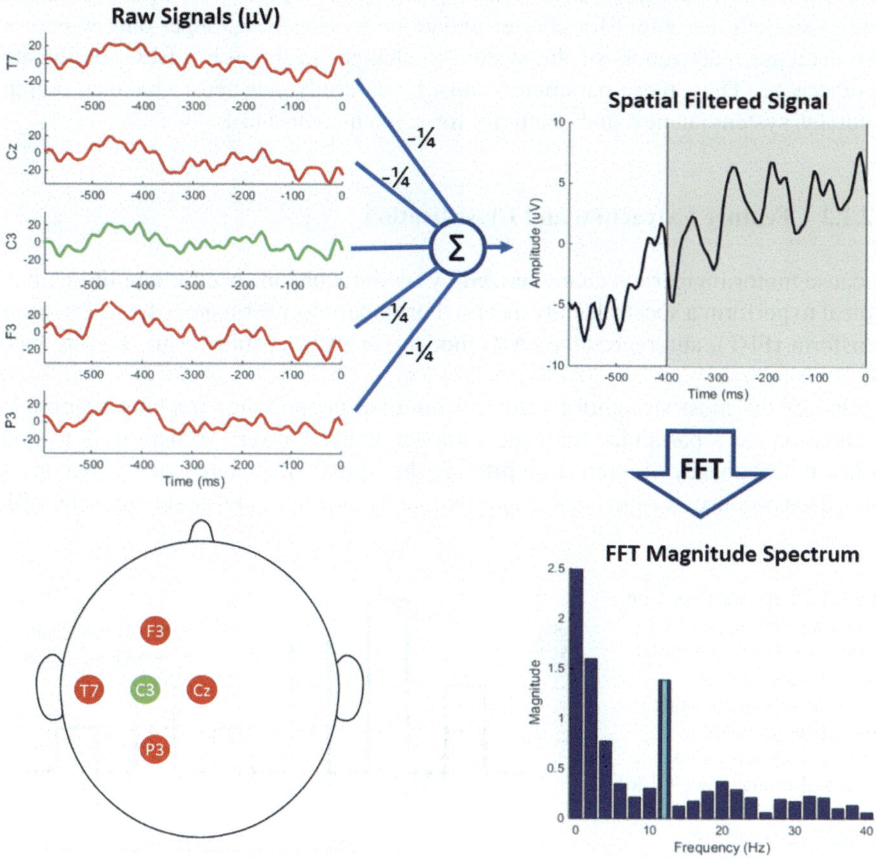

Fig. 9.2 Spectral analysis for a sensorimotor rhythm BCI. A Large-Laplacian spatial filter is applied centered on electrode C3, corresponding to the right-hand area of the sensorimotor cortex. The shaded region of the spatial filtered signal indicates a 400 ms data window. The FFT is computed with a 2-Hz resolution and the 12-Hz frequency bin corresponding to the rhythm is highlighted. This bin can be normalized and assigned to a 1-dimensional control variable with an update rate of 50 ms

The lower panel of Fig. 9.3, disregarding the stimulus labels from the upper panel, illustrates the windowing for asynchronous paradigms where the initial window onset begins with the data recording and does not correspond to any other external event. Subsequent windows are captured for analysis depending on the update rate, which can be as frequent as every signal sample. Assuming that a control command is issued for every data window, the update rate should be selected to minimize system output latency while accounting for the temporal dynamics of the control signal.

Similarly, the data window length should be selected to provide sufficient data for accurately classifying or translating the data while also accounting for the temporal dynamics of the control signal. The update rate and the data window also have an interaction. For example, a longer data window with respect to the update rate will tend to smooth the output for shorter update rates. However, longer data windows also decrease reactiveness of the system to changes in the user's EEG, deliberate or otherwise. Thus, these parameters must be carefully selected to balance output accuracy, system latency, and reactivity for a given control task.

9.2.1.2 Feature Extraction and Classification

Because motor imagery is characterized by modulations in spectral amplitude, it is typical to perform a spectral analysis based on bandpower estimates, the fast Fourier transform (FFT), autoregressive (AR) models, or wavelet transforms, for instance [4].

One of the most straightforward and intuitive methods for tracking amplitude modulations at a particular frequency, known as bandpower estimation, is to first isolate the frequency of interest by filtering the signal with a narrow-band bandpass filter. This produces a signal that is approximately an amplitude-modulated sinusoid.

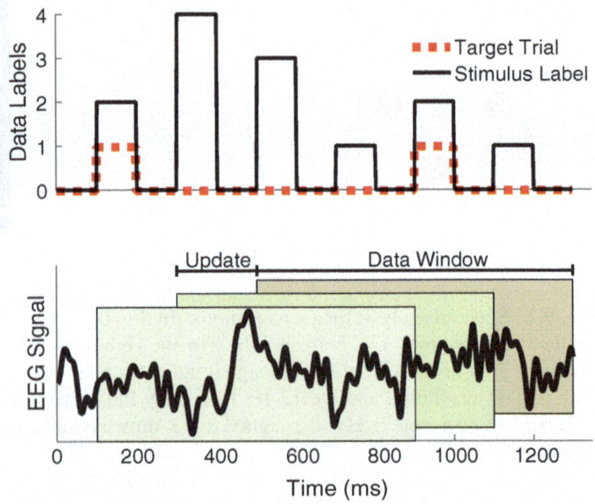

Fig. 9.3 Graphical depiction of data windowing for BCI processing. The upper panel shows the data labels corresponding to the sample EEG channel in the lower panel. For asynchronous designs, the data windows do not correspond to specific labeled events and are initiated at the beginning of the recording and updated according to the specified update rate

The signal is then rectified by squaring the signal or by computing its absolute value. The resulting rectified peaks are temporally smoothed together using a lowpass filter.

This process is illustrated in Fig. 9.4. Although the smoothed signal tracks the magnitude envelope of the frequency of interest, the resulting instantaneous magnitude estimate will be slightly delayed due to the filtering and smoothing steps. When multiple-frequency band tracking is required, it is generally more efficient to use an FFT- or AR-based method rather than using multiple bandpass filters and computing the band power of each output.

In contrast to bandpower estimation, the FFT provides the full frequency spectrum of the signal via a linear transform from the time to the frequency domain [28]. The FFT efficiently computes the discrete Fourier transform (DFT) given below:

$$X[k] = \sum_{n=0}^{N-1} x[n]e(-2\pi jnk/N), \tag{9.3}$$

where $x[n]$ is the time-domain signal, $X[k]$ is the frequency-domain representation, and N is the length of the DFT. Taking $x[n]$ as an EEG data window of length N, $X[k]$ will have N uniformly-spaced frequency bins between $\pm$(sampling rate)/2. To achieve frequency bins with different spacings, $x[n]$ can be zero-padded [28]. It is important to note that, while zero padding provides an interpolated spectrum, it does not increase the spectral resolution, which is limited to the length of the signal window before zero-padding (i.e., spectral resolution = sampling rate/the number of signal samples).

Computation of the DFT/FFT inherently causes spectral leakage, where the signal energy can "leak" into adjacent frequency bins. This leaked energy contributes to what are known as undesirable side lobes in the spectrum, flanking the main lobe of the desired signal energy. One approach to reducing spectral leakage is to multiply

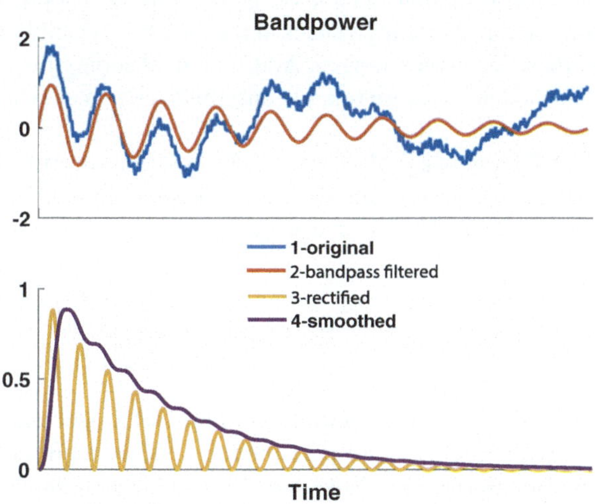

Fig. 9.4 The extraction of bandpower

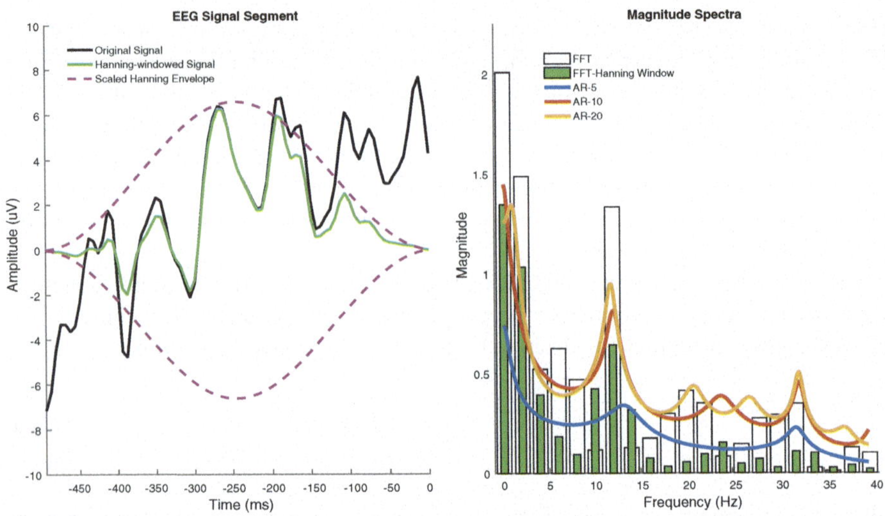

Fig. 9.5 Comparison of spectra generated by FFT, FFT with Hanning window, and AR models of 3 different orders. The left panel shows the time-domain signal before and after applying the Hanning window, also showing the shape of the Hanning window envelope, scaled for effect. The right panel shows The FFT of the original signal, the FFT of the Hanning-windowed signal, and spectra for 3 AR model orders using the original signal

the signal segment by a tapered window prior to the DFT computation as illustrated in Fig. 9.5. Note the side lobes that flank the 12 Hz peak of the regular FFT. While these side lobes are attenuated for the windowed FFT, it is observed that the main lobe around 12 Hz is broadened for the windowed FFT. There is a trade-off between main lobe width (i.e., spectral resolution) and side lobe-suppression that must be balanced based on the needs of the application. When tracking the amplitude of spectral peaks in standard frequency bands (e.g., μ, β) for BCI applications, windowing is generally preferred because high spectral resolution is typically not necessary for signal peaks in these frequency ranges. Additionally, keeping the signal energy in the main lobe tends to lead to more reliable amplitude estimates when accounting for signal noise.

AR models are also commonly used for spectrum estimation in BCI due to the fact that spectral resolution is not inherently limited by the length of the data window like the DFT [3]. The power spectrum for an AR model is given below:

$$\hat{P}_{\mathrm{AR}}(\omega) = \frac{\left|\hat{b}(0)\right|^2}{|1 + \sum_{k=1}^{P} \hat{a}_p(k)e^{-jk\omega}|^2}, \tag{9.4}$$

where $\hat{a}_p(k)$ are the coefficients of the AR model and p is the model order. Because the AR model is an all-pole model, it nicely represents peaks in the spectrum such as EEG oscillations. Note that the frequency variable ω, in contrast to the frequency

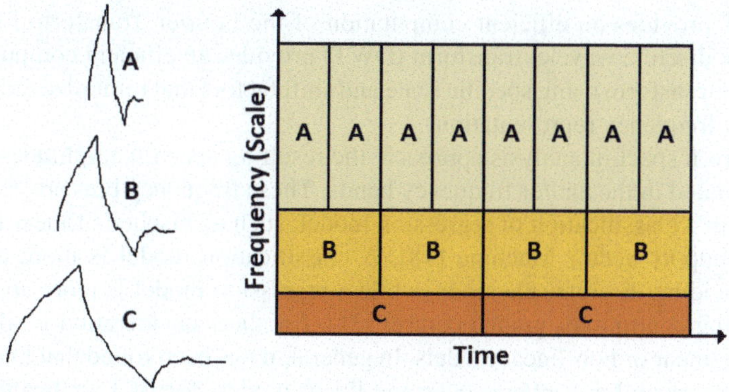

Fig. 9.6 The diagram on the right indicates the time-frequency tiling achieved by a wavelet analysis. Note that the higher-frequency content of a signal is computed over shorter time intervals (A), and the lower-frequency content is computed over longer time intervals (C). The time-domain waveforms on the left represent different scales of an example mother wavelet that could be used to compute the wavelet coefficient for the corresponding time-frequency tile

variable k in the DFT, is continuous. Thus, it is theoretically possible to estimate any frequency from a single model, although the practical frequency resolution is still fundamentally linked to the length of the data window. The trade-off with AR modeling is that the validity of the estimated spectrum depends on proper selection of the model order. Selecting a model order that is too low will result in an overly-smoothed spectrum, while a model order that is overestimated can create spurious peaks in the spectrum. See Fig. 9.5 for an example. Depending on types of filters applied in preprocessing and the dynamics of the signal for the intended application, common AR model orders for EEG range from 3 to 20, where the model order roughly approximates the number of spectral peaks to be captured [24].

Time-frequency approaches such as wavelet analysis aim to improve the balance between window length and spectral resolution [20, 29]. The concept is, rather than evaluating all frequencies over the same window size, high frequencies are evaluated over shorter windows while low frequencies are evaluated over longer windows. In wavelet analysis a characteristic time-limited pulse shape, called the mother wavelet, is correlated with the signal of interest at different time-shifts and time-scales. Since each scaled mother wavelet has a unique temporal length and represents a unique oscillation-frequency characteristic, the output of the correlation at each scale/shift represents a unique time-frequency component of the signal. This scheme results in a more effective, nonuniform time-frequency tiling compared to the FFT because changes in high-frequency characteristics can be identified over shorter time intervals than with the segment length used by the FFT. This time-frequency tiling and corresponding mother-wavelet scaling are illustrated in Fig. 9.6.

There are a wide variety of mother wavelets, and each has specific time-frequency characteristics and mathematical properties. In addition, application-specific mother wavelets can be developed if general pulse characteristics are known or desired. Just as the FFT provides an efficient computation of the Fourier Transform for digital signals, the discrete wavelet transform (DWT) provides an efficient computation of the wavelet transform using specific scale and shift factors that minimize redundancy in the time-frequency representation.

For a given spectral analysis approach, the resulting spectral amplitudes are typically evaluated in the and/or frequency bands. These frequency bins are commonly used to train a classification or regression model, such as Fischer's Linear Discriminant or Support Vector Machine [18]. A classification model is more appropriate for providing discrete selections, while a regression model is more appropriate for achieving continuous, graded control [31]. Another consideration is whether to implement linear or non-linear models. In general, it has been found that linear models can have several advantages over non-linear models for BCI applications [25]. Linear models are less prone to over-fitting with limited training data, and can offer added simplicity for computation, user training, and data interpretation.

Labeled events from a calibration session can be used to a train classification/regression model. The resulting classification will be performed on every sliding window, producing a continuous output if the update rate is sufficiently short. A block diagram of the preprocessing and feature extraction for sensorimotor rhythms is shown in Fig. 9.2. A similar approach can also be used to detect transient imagery events [21].

9.2.2 Common Spatial Patterns

A widely-used alternative to the traditional spectrum analysis approaches described in the previous section is the method of common spatial patterns (CSP) [26, 30]. CSP generates spatial filters that simultaneously minimize the variance for one class and maximize the variance for the other class, thus resulting in a simple classification based on the projected signal variances. An illustration of the CSP feature space for a 2-class scenario with 2 features is shown in Fig. 9.7. First, the signals are bandpass filtered in the range of interest. The CSP decomposition of a feature matrix is given as:

$$Y = WX, \tag{9.5}$$

where X is an N feature $\times$ T observance matrix, W is an L $\times$ N matrix (L $\leq$ N) whose L rows represent the individual components of the decomposition, and Y is an L $\times$ T matrix subspace of X. For a two-class classification problem, W can be determined to decompose the feature matrix such that the resulting projections corresponding to the extreme eigenvalues of the transformed covariance matrices have maximal variance for one class and minimal variance for the other class. First, for the two classes (1 and 2), the class-labeled observations are sorted by the respective class and the class-specific covariance matrices are determined:

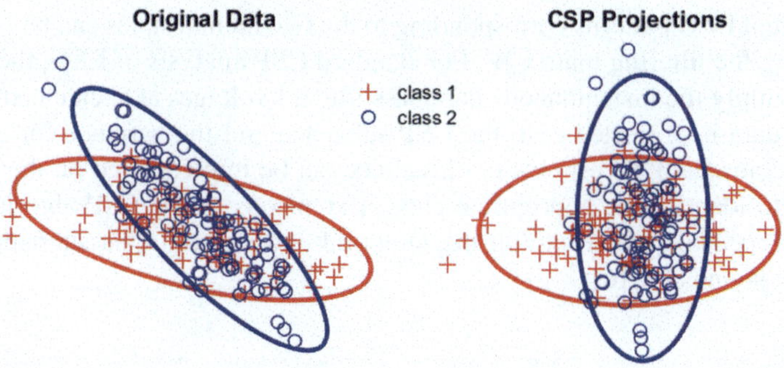

Fig. 9.7 Illustration of the CSP feature space for a 2-class scenario with 2 features. Note that, for a given orthogonal feature dimension, the CSP projections simultaneously minimize the variance for one class while maximizing the variance for the other class

$$\Sigma_1 = \mathrm{X}_{(1)}\mathrm{X}_{(1)}^{\mathrm{T}} \text{and } \Sigma_2 = \mathrm{X}_{(2)}\mathrm{X}_{(2)}^{\mathrm{T}}. \tag{9.6}$$

The task is defined as finding the transformation to create projections that simultaneously maximize the variance for one class and minimize the variance for the other:

$$W\Sigma_1 W^T = D \, and \, W\Sigma_2 W^T = I - D, \tag{9.7}$$

where D is a diagonal matrix with elements in [0,1]. This can be accomplished through simultaneous diagonalization of the two covariance matrices. First, a whitening transformation is performed:

$$P(\Sigma_1 + \Sigma_2)P^T = I. \tag{9.8}$$

Using spectral theory, the eigenvalue decomposition is then performed for the transformed classes:

$$P\Sigma_1 P^T = RDR^T \, and \, P\Sigma_2 P^T = R(I - D)R^T, \tag{9.9}$$

where the columns of are the eigenvectors and the diagonal elements of and are the eigenvalues of classes 1 and 2, respectively. Note that the maximum eigenvalues for one class correspond to the minimum eigenvalues for the other class. By selecting only the eigenvectors corresponding to the largest and smallest eigenvalues that provide the best discrimination between classes, the subspace projection matrix is defined as:

$$\tilde{W} = \tilde{R}^T P. \tag{9.10}$$

The actual EEG patterns corresponding to the two mental states can be visualized by inverting the filtering matrix W. For standard CSP analysis of EEG, the features of X are simply the instantaneous bandpass filtered voltages at each electrode. The incoming data is projected onto the CSP subspace and the variance for each projection is computed. Thresholds or classifiers can be implemented on the resulting variances to identify the appropriate class. Extensions of the CSP algorithm have been developed to further exploit the temporal and spectral characteristics of the underlying signals [9, 16].

9.3 Reactive Paradigms

9.3.1 Transient Evoked Potentials

Transient responses in reactive paradigms, known as evoked potentials (EPs), are time-locked to an external sensory stimulus. Thus, segments of EEG are analyzed over a predefined temporal window with respect to the onset of the stimulus. Because the signal-to-noise ratio (SNR) of EEG is low, multiple transient response observations are typically averaged to attenuate the background noise and produce a more reliable detection. While EPs can exhibit transient frequency bursts, they are most commonly analyzed using time-domain techniques. Thus, forms of spatio-temporal template matching are typically implemented to detect the relevant combination of amplitude deflections (e.g., voltages) at various channels that best characterize the response to the stimulus. The following describes a fundamental methodology for detecting and classifying EPs such as the P300 response [11].

9.3.1.1 Preprocessing

The appropriate channels must be identified for the EP to be analyzed, which are well-defined in the literature [2, 15]. Alternately, subspace decomposition approaches such as principle component analysis (PCA) [8] or independent component analysis (ICA) [12, 19] can be used to create a spatial filter that enhances the representation of the response for detection. Because EPs are generally comprised of low-frequency oscillations, it is common to bandpass filter the signals in the range of 0.1–0.5 Hz for the highpass cutoff and 10–30 Hz for the lowpass cutoff. It is also common to decimate the resulting signals based on the lowpass cutoff according to the Nyquist rate [28] to further reduce the dimensionality of the data.

9.3.1.2 Feature Extraction and Classification

The responses are collected based on the onset of each stimulus. Temporal windows for responses are typically around 1 s in length, but can vary depending on the application. Additionally, the window can begin prior to the onset of the stimulus to provide information about the baseline prior to the stimulus. Figure 9.3 illustrates the timing and associated EEG alignment of a typical synchronous stimulus presentation. The Stimulus Label in the upper panel (solid trace) indicates the onset and duration of each sensory stimulus. The lower panel shows the time-aligned EEG corresponding to the stimulus labels. The three shaded regions are example 800 ms windows corresponding to the onset of the first three stimuli in the upper panel. Similar response windows would be collected for all subsequent stimulus labels.

Assuming a binary detection scenario of predicting if the response was generated by a target or non-target stimulus, the collected EPs can be labeled and used to train a classifier [14]. For instance, in Fig. 9.3, all responses corresponding to Target Trial = 1 and all responses corresponding to Target Trial = 0 would be parsed for training the binary classifier. Figure 9.8 shows an example of the averaged P300 responses for target and non-target stimuli for the commonly-used electrode locations. Simple yet effective classifiers select the individual spatio-temporal features (circled) that optimize a regression model shown at the bottom. Using this approach, all features with high univariate correlation with the task are not necessarily selected for the model since they might have high covariance. Additionally, features with low univariate correlation may be included in the model to reduce noise or compensate for other selected features. This can also be generalized to multi-class problems. For classifying independent data, the resulting classifier scores are averaged over each stimulus label and the stimulus associated with the largest average score is classified as the selected target.

9.3.2 Steady-State Evoked Potentials

Steady-state responses such as steady-state visual evoked potentials (SSVEP) and steady-state somatosensory evoked potentials (SSSEP) in reactive paradigms generally present multiple, spatially-distinct stimuli, each at a unique frequency. Because the user focuses attention on a single stimulus (target) at a time in the presence of the other stimuli (distractors), the objective is to detect features of the EEG that are associated with the current target frequency. This forms a multiclass detection problem. Ideally, the EEG signal power at the target frequency and its harmonics will dominate compared to the distractor frequencies (and their respective harmonics) but this is not always the case for various reasons and more sophisticated techniques are employed to improve the detection.

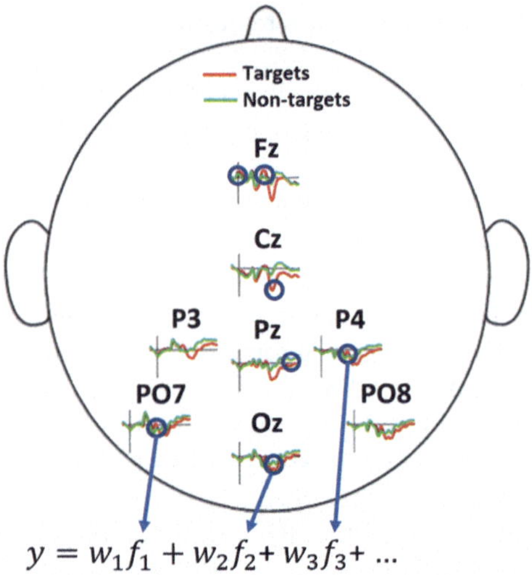

Fig. 9.8 Electrode locations and feature extraction for the visual P300 Speller. The averaged P300 ERPs for the target and non-target stimuli are shown for the commonly used electrode locations. Simple yet effective classifiers select the individual spatio-temporal features (circled) that optimize a regression model shown at the bottom. Using this approach, all features with high univariate correlation with the task are not necessarily selected for the model since they might have high covariance. Additionally, features with low univariate correlation may be included in the model to reduce noise or compensate for other selected features

9.3.2.1 Preprocessing

The appropriate channels must be identified for the response to be analyzed (e.g., SSVEP, SSSEP), which are well-defined in the literature [1, 13, 27]. The EEG can be bandpass filtered between 0.1 Hz (or just below the lowest stimulation frequency) and just above the frequency of the maximum stimulus harmonic of interest to eliminate noise outside of the frequency range of interest. Similar to Sect. 2.2.1, the EEG is evaluated continuously using overlapping windows. In this case the window length typically ranges from 1 to 2 s with a 0.5 s update rate.

9.3.2.2 Feature Extraction and Classification

While spectral analysis approaches similar to those discussed in Sect. 9.2.1 can be used to detect and classify SSVEP responses, the most widely accepted algorithm for such stimuli is canonical correlation analysis (CCA) [6, 17]. CCA is a multi-dimensional statistical analysis technique that finds underlying linear correlations between two sets of data. For BCI, the CCA algorithm effectively generates a spatial filter associated with each target frequency that produces the highest correlation for a given data window. The spatial filter that produces the highest correlation designates the current target of the user's attention. This approach has an advantage over standard spectral analysis, e.g., techniques based on the Fourier Transform, in that it simultaneously combines spatial and spectral information in the classification decision and tends to provide more reliable performance. This approach does not require calibration or training prior to online operation and allows for continuous, asynchronous operation.

Given two multi-dimensional data sets, X and Y, linear combinations $x = X^T W_x$ and $y = Y^T W_y$ can be found that maximize the correlation between x and y. CCA finds the weight vectors W_x and W_y by solving the following optimization problem:

$$\max_{W_x, W_y} \frac{E\left[W_x^T X Y^T W_y\right]}{\sqrt{\left(E\left[W_x^T X X^T W_x\right] E\left[W_y^T Y Y^T W_y\right]\right)}}. \tag{9.11}$$

In practice, this can be solved using the singular-value decomposition method to diagonalize the covariance matrices as the maximum canonical correlation corresponds to the square-root of the largest eigenvalue.

For BCI, CCA generates a spatial filter for multichannel EEG data, X, that maximizes the correlation between a set of sinusoidal templates Y_f at each target frequency. This reference set consists of sine and cosine signals at the fundamental and harmonic frequencies of each stimulus, and results in reference waveforms that match the temporal length of the EEG window. The idea is that the sinusoidal templates corresponding to the target frequency should better match the EEG than the templates at the other frequencies. The reference signal Y_f (9.12) can be derived using N_h harmonics, where f is the fundamental frequency and t is time.

$$\mathrm{Y_f} = \begin{pmatrix} \sin(2\pi f t) \\ \cos(2\pi f t) \\ \\ \vdots \\ \\ \sin(2\pi N_h f t) \\ \cos(2\pi N_h f t) \end{pmatrix}. \tag{9.12}$$

EEG data is canonically correlated with each reference signal and the classification output is determined as $f_s = \max_i \rho(f)$, where $f = f_1, f_2, \ldots f_k$ and K is the total number of classes (target frequencies) in the BCI. See Fig. 9.9 for a graphical depiction of CCA.

Both sine and cosine templates are used because a linear combination of the two can represent a single sinusoid with arbitrary phase, matching the characteristics of the EEG observation. Typically, only 2 or 3 harmonics are needed for an accurate classification, but the number of harmonics can be easily reconfigured to meet the performance needs. For each EEG data segment, CCA is performed for the template corresponding to each target frequency. CCA returns a set of optimized spatial weights for the EEG channels and the CCA sinusoidal templates that maximize the

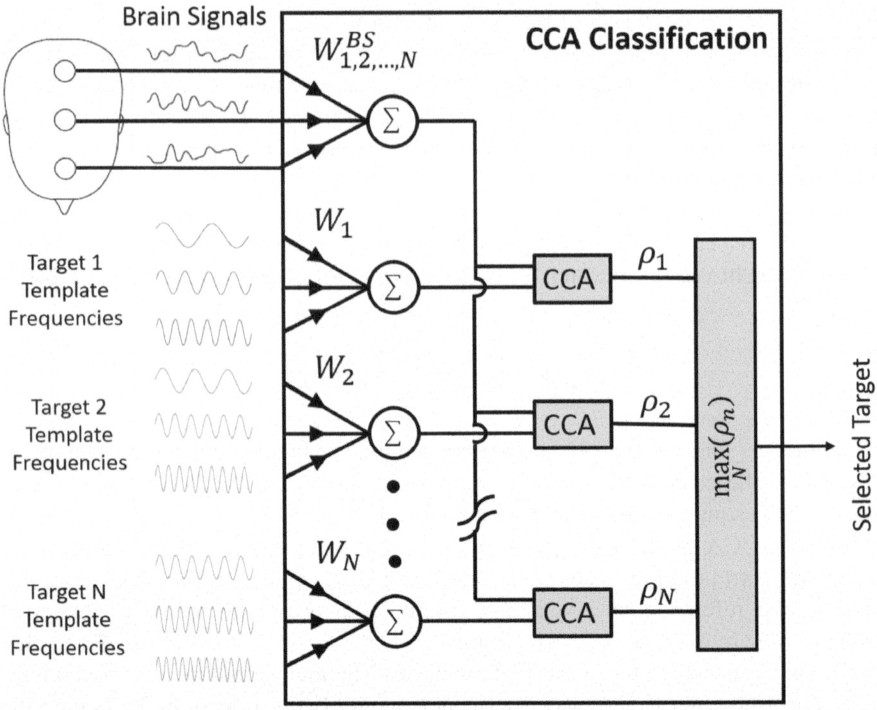

Fig. 9.9 Graphical depiction of canonical correlation analysis (CCA) for an n-class SSVEP paradigm. For each of N targets, a weighted sum of EEG channels is correlated with a weighted sum of sinusoidal templates at the harmonic frequencies (3 in this case) of the respective target stimulus. The optimal weights are computed separately for each target via CCA, which produces a maximized Pearson correlation coefficient. The resulting correlation coefficients are compared across targets. The target that produced the maximum correlation is output as the current selection. Note that distinct EEG signal weights are generated for each target, corresponding to the subscript of the weight matrix

resulting correlation, as well as the value of this correlation. The target frequency corresponding to the CCA template that produces the largest correlation for the given data segment is selected as the output. This process is repeated for each subsequent EEG data window for asynchronous operation. Additionally, null-state detection can be implemented with an appropriate threshold on the correlation values. Similar and more sophisticated approaches have been developed for code-modulated visual evoked potentials [5] and very high information transfer rate SSVEP paradigms [7].

9.4 Other Considerations

9.4.1 *Artifact Removal*

When needed, BCIs implement many of the same artifact removal techniques discussed in Chap. 2. However, for certain BCI applications, it may not be worth the additional effort and computational resources to design and implement artifact removal for rare or inconsequential artifacts such as eye blinks, for instance. It is common to design processing and classification stages to be tuned to the control signal of interest and inherently immune to artifacts. For many of the aforementioned BCI paradigms, the control signals have a distinct spatial and spectral distribution from common artifacts, which can be sufficiently attenuated via standard spatial or spectral filtering without the need for specialized artifact characterization or processing. Without careful design, an artifact removal technique may also further distort the control signal and increase computational resources such that there is ultimately little or no practical benefit over excluding artifact removal from the processing chain.

9.4.2 *Real-Time Processing*

With the considerable advances available in modern computing technology, the real-time processing requirements for BCIs are no longer as constraining as they once were. All of the methods presented in this chapter are fully able to be implemented with minimal delays due to signal processing.

Most standard signal preprocessing steps such as spatial and spectral filtering can be efficiently implemented to achieve real-time feedback. For instance, spatial filter parameters can be computed offline with online implementations that require negligible computational resources. Online computation of spatial filters such as online ICA requires significant computational resources for real-time feedback and are not commonly implemented for BCIs. For spectral filtering, infinite impulse response (IIR) filter structures are preferred over finite impulse response (FIR) structures. For causal online filtering, symmetric FIR filters introduce a delay equivalent to half the filter length, while IIR filters with comparable frequency characteristics can be designed having significantly shorter latencies. The trade-off for IIR filters compared to FIR is that they introduce phase distortion in signals. Depending on the application, this may be tolerable or included in the classifier design without detriment. If phase distortion is not tolerable, zero-phase filtering can be implemented to eliminate the phase distortion while effectively doubling the latency compared to using the equivalent filter without a zero-phase implementation [28].

However, it should still be noted that BCIs based on continuous responses are often far more restrictive than those based on transient responses. For instance, consider a continuous BCI using SSVEP. The processing should occur within a single data window in order to prevent a lagged control feedback to the user, which could prove distracting and detrimental to usability and performance. If many stimulus frequencies are used, sufficient computational resources are needed to maximize the correlation between the sinusoidal templates and the EEG to avoid an appreciable lag in the feedback.

9.4.3 User Versus System Adaptation

For generally stable responses such as P300 and SSVEP, BCI designs with static processing and classification are typically adequate. For some paradigms such as motor imagery, the user can learn to better focus and modulate brain responses via training [33]. Additionally, the brain state and background brain activity of the user can change over time, even within a session. Thus, static designs where the user is forced to adapt to the BCI feedback can be suboptimal. Alternately, the BCI can be designed to adapt its processing and classification to the user's changes in brain activity and/or performance. The challenge is that this system adaptation must be done using periodic calibration sessions, or without calibration sessions in an unsupervised or semi-supervised manner. Furthermore, it is imperative to select an appropriate rate of adaptation, which can be user-dependent [23]. Ultimately, the implementation of an adaptive BCI results in a co-adaptive system since the user will inevitably adapt to the provided feedback. Co-adaptive systems can be highly prone to instability and it is vital to carefully design and select the adaptation parameters.

References

1. S. Ahn, K. Kim, S.C. Jun, Steady-state somatosensory evoked potential for brain-computer interface-present and future. Front. Hum. Neurosci **16**, 832 (2015)
2. F. Aloise, I. Lasorsa, F. Schettini, A. Brouwer, D. Mattila, F. Babiloni, F. Cincotti, Multimodal stimulation for a P300-based BCI. Int. J. Bioelectromagn. **9**, 128–130 (2007)
3. C.W. Anderson, E.A. Stolz, S. Shamsunder, Multivariate autoregressive models for classification of spontaneous electroencephalographic signals during mental tasks. IEEE Trans. Biomed. Eng. **45**, 277–286 (1998)
4. A. Bashashati, M. Fatourechi, R.K. Ward, G.E. Birch, A survey of signal processing algorithms in brain–computer interfaces based on electrical brain signals. J. Neural Eng. **4**, R32–R57 (2007)
5. G. Bin, X. Gao, Y. Wang, Y. Li, B. Hong, S. Gao, A high-speed BCI based on code modulation VEP. J. Neural Eng. **8**, 025015 (2011)
6. G. Bin, X. Gao, Z. Yan, B. Hong, S. Gao, An online multi-channel SSVEP-based brain—computer interface using a canonical correlation analysis method. J. Neural Eng. **6**, 046002 (2009)
7. X. Chen, Y. Wang, M. Nakanishi, X. Gao, T.P. Jung, S. Gao, High-speed spelling with a noninvasive brain–computer interface. Proc. Natl. Acad. Sci. USA **112**, E6058–E6067 (2015)
8. J. Dien, K.M. Spencer, E. Donchin, Localization of the event-related potential novelty response as defined by principal components analysis. Cognitive Brain Res. **17**, 637–650 (2003)
9. G. Dornhege, B. Blankertz, M. Krauledat, F. Losch, G. Curio, K.R. Muller, Combined optimization of spatial and temporal filters for improving brain-computer interfacing. IEEE Trans. Biomed. Eng. **53**, 2274–2281 (2006)

10. L.A. Farwell, E. Donchin, Talking off the top of your head: toward a mental prosthesis utilizing event-related brain potentials. Electroencephalogr. Clin. Neurophysiol. **70**(6), 510–523 (1988)
11. R. Fazel-Rezai, B.Z. Allison, C. Guger, E.W. Sellers, S.C. Kleih, A. Kübler, P300 brain computer interface: current challenges and emerging trends. Front. Neuroeng. **5**, 14 (2012)
12. A. Hyvärinen, E. Oja, Independent component analysis: algorithms and applications. Neural. Netw. **13**, 411–430 (2000)
13. S.P. Kelly, E.C. Lalor, R.B. Reilly, J.J. Foxe, Visual spatial attention tracking using high-density SSVEP data for independent brain-computer communication. IEEE Trans. Neural Syst. Rehabil. Eng. **13**, 172–178 (2005)
14. D.J. Krusienski, E.W. Sellers, F. Cabestaing, S. Bayoudh, D.J. McFarland, T.M. Vaughan, J.R. Wolpaw, A comparison of classification techniques for the P300 Speller. J. Neural Eng. **3**, 299 (2006)
15. D.J. Krusienski, E.W. Sellers, D.J. McFarland, T.M. Vaughan, J.R. Wolpaw, Toward enhanced P300 speller performance. J. Neurosci. Methods **167**, 15–21 (2008)
16. S. Lemm, B. Blankertz, G. Curio, K.R. Muller, Spatio-spectral filters for improving the classification of single trial EEG. IEEE Trans. Biomed. Eng. **52**, 1541–1548 (2005)
17. Z. Lin, C. Zhang, W. Wu, X. Gao, Frequency recognition based on canonical correlation analysis for SSVEP-based BCIs. IEEE Trans. Biomed. Eng. **54**, 1172–1176 (2007)
18. F. Lotte, M. Congedo, A. Lécuyer, F. Lamarche, B. Arnaldi, A review of classification algorithms for EEG-based brain–computer interfaces. J. Neural Eng. **4**, R1 (2007)
19. S. Makeig, S. Enghoff, T.P. Jung, T.J. Sejnowski, A natural basis for efficient brain-actuated control. IEEE Trans. Neural Syst. Rehabil. Eng. **8**, 208–211 (2000)
20. S. Mallat, *A Wavelet Tour of Signal Processing* (Academic Press, Orlando, FL, 2008)
21. D.J. McFarland, D.J. Krusienski, W.A. Sarnacki, J.R. Wolpaw, Emulation of computer mouse control with a noninvasive brain–computer interface. J. Neural Eng. **5**, 101 (2008)
22. D.J. McFarland, L.M. McCane, S.V. David, J.R. Wolpaw, Spatial filter selection for EEG-based communication. Electroencephalogr. Clin. Neurophysiol. **103**, 386–394 (1997)
23. D.J. McFarland, W.A. Sarnacki, J.R. Wolpaw, Should the parameters of a BCI translation algorithm be continually adapted? J. Neurosci. Methods **199**, 103–107 (2011)
24. D.J. McFarland, J.R. Wolpaw, Sensorimotor rhythm-based brain–computer interface (BCI): model order selection for autoregressive spectral analysis. J. Neural Eng. **5**, 155 (2008)
25. K.R. Muller, C.W. Anderson, G.E. Birch, Linear and nonlinear methods for brain-computer interfaces. IEEE Trans. Neural Syst. Rehabil. Eng. **11**, 165–169 (2003)
26. J. Müller-Gerking, G. Pfurtscheller, H. Flyvbjerg, Designing optimal spatial filters for single-trial EEG classification in a movement task. Clin. Neurophysiol. **110**, 787–798 (1999)
27. A.M. Norcia, L.G. Appelbaum, J.M. Ales, B.R. Cottereau, B. Rossion, The steady-state visual evoked potential in vision research: a review. J. Vis. **15**, 4 (2015)
28. J. Proakis, D. Manolakis, *Digital Signal Processing: Principles, Algorithms and Applications* (Prentice Hall, New York, NY, US, 2007)
29. L. Qin, B. He, A wavelet-based time–frequency analysis approach for classification of motor imagery for brain–computer interface applications. J. Neural Eng. **2**, 65 (2005)
30. H. Ramoser, J. Muller-Gerking, G. Pfurtscheller, Optimal spatial filtering of single trial EEG during imagined hand movement. IEEE Trans. Rehabil. Eng. **8**, 441–446 (2000)
31. E.W. Sellers, D. Krusienski, D. Mcfarland, J. Wolpaw, Non-invasive brain-computer interface research at the wadsworth center, in *Toward Brain-Computer Interfacing*, ed. by G. Dornhege, J.R. Millan, T. Hinterberger, D.J. McFarland, K.R. Muller (The MIT Press, Cambridge, 2007), pp. 31–42
32. J.R. Wolpaw, D.J. McFarland, Multichannel EEG-based brain-computer communication. Electroencephalogr. Clin. Neurophysiol. **90**, 444–449 (1994)
33. J.R. Wolpaw, D.J. McFarland, Control of a two-dimensional movement signal by a noninvasive brain-computer interface in humans. Proc. Natl. Acad. Sci. USA **101**, 17849–17854 (2004)

34. T.O. Zander, C. Kothe, S. Jatzev, M. Gaertner, Enhancing human-computer interaction with input from active and passive brain-computer interfaces, in *Brain-Computer Interfaces*, ed. by D.S. Tan, A. Nijholt (Springer, London, 2010), pp. 181–199
35. T.O. Zander, C. Kothe, Towards passive brain–computer interfaces: applying brain–computer interface technology to human–machine systems in general. J. Neural Eng. **8**, 025005 (2011)

Chapter 10
Computational EEG Analysis for Hyperscanning and Social Neuroscience

Dan Zhang

Abstract Hyperscanning, the technique that simultaneously records neural activities from multiple interacting participants, has attracted increasing attention in the field of social neuroscience. EEG is among the most popular neuroimaging techniques for hyperscanning, as its high portability enables neural signal recordings in naturalistic social interaction scenarios. This chapter summarizes the state-of-the-art progress on the computational EEG analysis methods for hyperscanning and social neuroscience. These methods are divided into two categories, focusing on social perception and social interaction, respectively. A variety of computational models have been proposed and implemented to quantitatively describe the hyperlinks among interacting brains, and significant hyperlinks have been reported in social tasks covering typical social activities. As the development of hyperscanning methods is still at its early beginning, future perspectives are discussed at the end of the chapter.

10.1 Introduction

Humans are fundamentally a social species, rather than individualists. Social activities are hereby essential for humans. With the rapid development of neuroscientific research techniques, increasing interest has been drawn toward social neuroscience, which is an interdisciplinary field devoted to the understanding of biological implementation of social processes and behaviors [37, 60]. To explore the neural basis of social behaviors, the conventional single-brain approach has been widely used and great progresses have been made. Based on data collected from both patients with social function disorders and healthy people, the important brain regions for social functioning, such as the fusiform area for perceiving facial information, the Broca's and the Wernicke's areas for processing speech information, the mirror neuron network for interpreting actions, etc., have been identified and theories about their working mechanisms have been (partially) elucidated [19, 65]. Despite its great

D. Zhang (✉)
Department of Psychology, School of Social Sciences, Tsinghua University, Beijing, China
e-mail: dzhang@tsinghua.edu.cn

C.-H. Im (ed.), *Computational EEG Analysis*, Biological and Medical Physics, Biomedical Engineering, https://doi.org/10.1007/978-981-13-0908-3_10

success, the single-brain approach has recently been criticized, mostly for its limitations on ecological validity [8, 32, 38, 49]: the experimental paradigms of the single-brain approach often employ 'off-line' social cognition tasks with abstract stimuli, resulting in reduced generalizability of the findings. Although paradigms of high ecological validity are always preferred, naturalistic social stimuli and real-time online social interactions pose great challenges for developing effective research methods.

To further pursue neural mechanisms of our social brain, the hyperscanning technique has been developed and applied for neuroscience research in the past two decades [2, 14, 23]. The hyperscanning technique aims at collecting and analyzing neural signals from multiple persons involved in naturalistic social scenarios. The collection of neural signals from interacting brains is believed to overcome the major critiques on the single-brain approach: our brains should be better activated during real social activities, rather than in isolated single-brain paradigms. To date, a variety of brain imaging techniques has been employed in hyperscanning studies. The very first study in 1965 by Duane and Behrendt utilized dual-EEG to explore a possible extrasensory electroencephalographic induction between identical twins [12]. The modern hyperscanning studies started from an fMRI study by Montague and colleagues, demonstrating a hyperlink between two persons playing a simple deception game in simultaneously recording fMRI scanners [39]. Since then, more than 80 papers have been published (source: PubMed), using fMRI, EEG or fNIRS. While fMRI has the best spatial resolution, EEG and fNIRS has gained increasing popularity in recent years, for their high portability and low running cost. Compared to both fMRI and fNIRS, EEG has unique advantages for its rich temporal and spectral information. More importantly, the millisecond-scale high temporal resolution of EEG is capable of following the fast temporal dynamics of human social activities. Nevertheless, further development of computational EEG analysis methods is required, as EEG-based interacting-brain approach emphasizes the interplay among multiple persons, for which the conventional single-brain based analysis methods cannot be directly applied. While substantial progress has been made toward hyperscanning-specific analysis methods in the past decade, we are still at the early beginning of exploring the 'hyperlinks' among interacting brains.

In this chapter, we review recent works on computational EEG analysis methods for social neuroscience studies using the interacting-brain approach (by employing the EEG-based hyperscanning technique). Our review is further divided into two sessions, 'social perception' and 'social interaction'. Whereas the 'social perception' section introduces these methods on perceiving naturalistic social information, the 'social interaction' section focuses on methods for characterizing the inter-person social activities.

10.2 Hyperlink for Social Perception

Perception of social information is the foundation for human social behaviors. Our brain has been shown to have highly specified regions for efficiently processing social information, such as faces and voices [60]. The state-of-the-art findings, however, are mostly conducted using single-brain approaches and the event-related method has been the most popular and most widely-used technique. Due to the complicated and strong noise in neural signals, the event-related method normally requires a highly controlled experimental environment, with a sufficient number of repetition of the target social stimuli. These requirements, however, lead to experimental setups that are quite different from real-world naturalistic social scenarios. In addition, neural responses in the event-related method are defined by activities significantly different from a certain pre-selected baseline that is expected to be free of stimulus-related brain activities. Due to the non-stationarity nature of neural activities [28], these baselines are suggested to be temporally close to the time of the stimuli, e.g. up to several hundred milliseconds prior to the onset of the stimuli for EEGs. The stimulus-free baseline periods, again, is not likely to be feasible in real-world situations as people are always involved in continuous perception of social information.

Calculating the hyperlink among a group of people perceiving the same social stimuli, is proposed to be a promising tool to explore the neural responses during social perception (Fig. 10.1). In contrast to the conventional event-related methods, the hyperlink methods define social relevant neural responses as activities showing significant similarities across participants. Hyperlink works on the basis of the reliability of neural responses, hereby avoids the challenging issue of defining events in the complex and naturalistic social stimulation streams. Although simultaneous

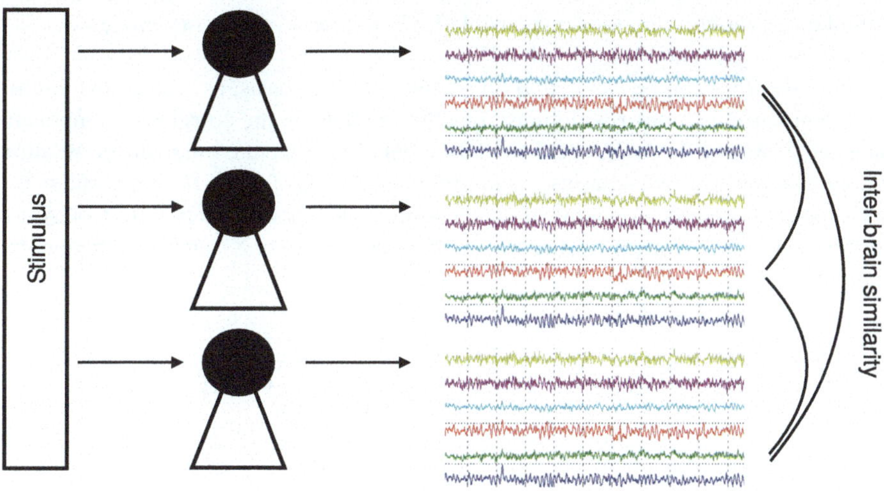

Fig. 10.1 Illustration of the methodology for calculating hyperlink for social perception

recording of multiple participants is preferred, this kind of hyperlink can be applied to 'off-line' recorded data as well, as long as the data are from participants receiving identical social stimulations.

Indeed, a number of studies have shown that human brain activities can be highly reliable under naturalistic stimulus conditions. Originally termed as inter-subject correlation (ISC) and applied for fMRI data, significant hyperlinks were found when a group of five participants watched half an hour of a popular movie. Results revealed that hyperlinks existed beyond the primary and secondary visual cortex, including higher-level visual processing regions, auditory regions, etc. [21]. Follow-up fMRI studies reported that ISC spatial patterns could be modulated by the content of the visual stimuli, with unstructured movie clips showing the minimal hyperlink-based activation and highly structured movie clips eliciting a reliable hyperlink across a widespread brain network including the parietal and frontal regions as well [22]. Another ISC-based fMRI study reported a bilateral network for speech production, extending our previous understanding of left lateralized network [57]. The calculated hyperlink could also predict human behaviors, for instance, the powerfulness of political speeches [51]. The hyperlink in fMRI data is usually calculated using the inter-subject correlation method, which is the average of all pair-wise Pearson correlations for each individual voxel, as follows:

$$\mathrm{ISC} = \frac{1}{m(m-1)/2}\sum_{i=1}^{m}\sum_{j=2,j>i}^{m} r_{ij} \tag{10.1}$$

where m is the number of participants, and $\mathrm{r_{ij}}$ represents the temporal (Pearson's) correlation between participant i and j, given a specific voxel or channel. Such a pairwise correlation based ISC calculation has also been applied for EEG data analysis. Reliable pairwise ISCs have been observed when a variety of different naturalistic social stimulations for brain regions responsible for both low-level sensory processing and high-level social functioning [3, 4, 27].

Pairwise ISCs can be calculated on a single-channel basis, as well as in a multivariate manner. The most widely used method to date, is the correlated component analysis (CoCA) [9]. CoCA seeks to find spatial filters that maximize the correlation among two multivariate datasets. As multi-channel EEGs from different participants are supposed to perceive identical social stimuli, the spatial filters w in CoCA are formulated to be identical for the two multivariate datasets $\mathrm{X_1}$ and $\mathrm{X_2}$ (channel by sample)

$$\begin{aligned} max\ Corr\left(w^T X_1, w^T X_2\right) &= max\frac{w^T X_1 \cdot \left(w^T X_2\right)^T}{\left|w^T X_1\right| \cdot \left|w^T X_2\right|} \\ &= max\frac{w^T R_{12} w}{\sqrt{w^T R_{11} w} \cdot \sqrt{w^T R_{22} w}} \end{aligned} \tag{10.2}$$

where R_{11}, R_{22}, and R_{12} represent the covariance matrices

$$R_{ij} = X_i \cdot X_j^T . \tag{10.3}$$

The above spatial filters w can be obtained via a generalized eigenvalue decomposition based on solving the following optimization problem [17, 44]:

$$\lambda \cdot (R_{11} + R_{22}) \cdot w = R_{12} \cdot w . \tag{10.4}$$

Although the above formula is originally designed for two multivariate datasets, Dmochowski and colleagues proposed to construct two datasets that include all unique combination of pairs of participants [9]. A three-participant version is shown below

$$\begin{aligned} X_1 &= [X_{P1} X_{P1} X_{P2}], \\ X_2 &= [X_{P2} X_{P3} X_{P3}], \end{aligned} \tag{10.5}$$

where X_{Pi} represents multi-channel EEG data (channel by sample) from the i-th participant. By employing such a multivariate construct, the CoCA method hereby can obtain the spatial filters that maximize the 'correlation' among EEG data from all participants. It is worthwhile to note that the number of extracted CoCA components (each corresponding to a specific spatial filter) are normally substantially reduced compared to the number of original EEG channels, therefore the manual efforts needed for further data analysis can be greatly reduced. More importantly, as the optimization problem is defined based on a relatively simple and straightforward assumption about reliability across participants, the obtained spatial filters find neurophysiologically plausible components representing shared neural activities. Indeed, the extracted components have been reported to be specifically responsive to certain social stimuli, including face, hand, or high-level social emotions such as surprise, tension, anticipation, etc. [10, 30, 69].

Beside the exploration in the spatial domain, inter-brain hyperlink has also been investigated in the spectral and temporal domains as well, but to a less extent. Neural oscillations at different frequency bands such as alpha, theta, delta, etc., have long been known to have important functional roles for human cognition [52, 56]. It is hereby reasonable to assume that inter-brain hyperlink may rely on oscillatory brain activities. Although exploration in this direction are just beginning, it has been reported that hyperlink in the delta band had primary contribution to behaviorally measured audience preference [3]. In the time domain, the most critical question is the optimal time window length for calculating hyperlink. Due to the non-stationarity of EEG signals, calculating temporal correlation with a long time window may lead to unstable results. To date, the reported studies have used time windows ranging from 200 ms (e.g. [4] up to several minutes (e.g. [3, 10]. Although time window as short as 200 ms has been demonstrated to be capable of capturing neural reliability, hyperlinks based on different time window lengths may have different functional implications, which can be considered as the counterpart for the spectral domain analysis. In addition, dissociation of EEG temporal signals into phase and amplitude may worth further exploration, as they have long been postulated to have distinct roles for a variety of human cognition functions [48, 68, 70], hereby it is reasonable to expect different functional roles for hyperlinks as well.

Another important aspect to be explored is the number of participants for reliable hyperlink estimation. While such a question has been addressed using fMRI and 20 is suggested to be the minimal requirement [43], a systematical investigation for EEG-based hyperlink is lacking. The number of participants employed in EEG-based studies ranged from 4 to approximately 20. Further studies with a substantially increased number of participants (preferably to be offline collections) are necessary to evaluate the suitable number of participants.

10.3 Hyperlink for Social Interaction

Interaction with other human beings, is the essential building element of our society. Studies on hyperlink for social interaction normally demand simultaneous recording of multiple brains, with only a few exceptions, in which the researchers were interested in uni-directional information transfer (e.g. [61]. Consequently, here different interacting participants are likely to have different roles and their brain activities cannot be simply considered to comprise similar information (Fig. 10.2).

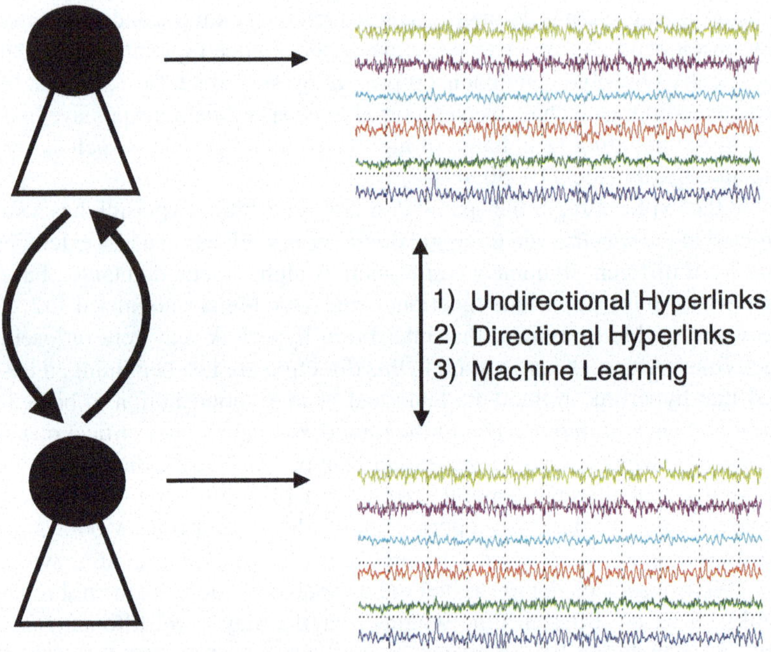

Fig. 10.2 Illustration of the methodology for calculating hyperlink for social interaction

Hereby, more advanced data analysis methods are required to describe the dynamic interaction across different brains.

The very first hyperlink study on social interaction was also conducted using fMRI. Montague and colleagues reported a significant correlation in supplementary motor areas between two persons playing a deception game [39]. Their further and more influential work revealed an fMRI-based neuro-link between two players in a trust game that could predict the intention to trust [31]. Speech, as a dominant social interaction type, has also been studied using fMRI-based hyperscanning. The dynamic neural coupling between a speaker and a listener was shown to have different patterns in different brain regions: while the auditory related regions exhibited a delayed coupling of the listener as compared to the speaker, the frontal regions revealed a leading coupling by the listener possibly responsible for speech anticipation [61]. Furthermore, it has been suggested that interactive experience and skills play enabling roles in social cognitive functions [7, 8]. Hyperlink hereby provides a promising new perspective for probing the interactive nature of our brain.

The first EEG hyperscanning study for social interaction by Babiloni and colleagues involved sets of four individuals playing Tressette, a bridge-like game (Babiloni et al. [1]. The portability of EEG has enabled researchers to extend the exploration into more realistic conditions. To date, researchers have recorded multi-person EEGs in scenarios representing the majority of our social interaction activities, ranging from simple motor actions to complex activities such as making a conversation, performing music, playing games, etc. [2]. As different participants may have different roles, the information flow among participants may have different directions, with some participants leading the others. The methods for characterizing hyperlinks can be summarized into the following three categories: (1) undirectional hyperlink methods; (2) directional hyperlink methods; (3) machine-learning methods.

10.3.1 Undirectional Hyperlink Methods

This category of methods is straightforwardly derived from the inter-subject correlational methods as used for describing social perception. The fundamental hypothesis underlying most of the correlational methods is that different brains have similar and synchronized time courses of neural activities, therefore suitable for the cases when participants have balanced roles. Pairwise Pearson's correlation can be calculated for EEG signals filtered at different frequency bands. Inter-brain correlation of the amplitudes of theta and alpha over right temporal-parietal junction (TPJ), the amplitudes of alpha and beta over frontal regions, have been reported to be associated with the understanding of others' intention and high-level cooperative strategies [29, 58]. Phase-locking analysis is another type of method that has been frequently employed. Phase-locking analysis focuses on the circular correlation of the phases of the neural oscillations, usually termed as the phase-locking value (PLV), as follows:

$$\mathrm{PLV(t)} = \frac{\left|\sum_{i=1}^{N} \exp(j\varphi_i(t))\right|}{\mathrm{N}}, \tag{10.6}$$

where $\varphi_i(t)$ is the phase of the neural oscillation at a certain frequency band for the i-th participant at time point t, and N is the number of participants. PLV is a measure independent of amplitude fluctuation: a high inter-brain PLV implies a more synchronized pace among the participants' neural activities. Significant inter-brain PLVs are frequently observed in relatively low frequency bands, such as delta, theta, alpha, covering a variety of social interaction paradigms [13, 18, 26, 41, 45, 62].

The correlational methods can be extended to a multivariate version as well, by incorporating the conventional canonical correlation analysis (CCA) method, as follows:

$$max\ Corr\left(w_1^T X_1, w_2^T X_2\right) = max \frac{w_1^T X_1 \cdot \left(w_2^T X_2\right)^T}{\left|w_1^T X_1\right| \cdot \left|w_2^T X_2\right|}, \tag{10.7}$$

where X_1 and X_2 are two multivariate EEG datasets and w_1 and w_2 are two to-be-calculated spatial filters that maximize the linear correlations between the two datasets. The optimization problem is similar to CoCA [see (10.2)] but the spatial filters are allowed to be different for the two datasets, thus supporting different roles for different brains. However, we only found one study that utilized CCA to measure the correlation between listeners' and speakers' EEGs and the authors reported an attentional modulation of the listener-speaker hyperlink [35]. Nevertheless, multivariate analysis methods are a necessary extension of the present univariate methods for characterizing the complex inter-brain coupling during social interaction. CCA for multiple datasets (i.e. more than two datasets) is a promising candidate for further exploration, as it fits very well with the nature of the multi-brain design [54, 68, 70].

10.3.2 Directional Hyperlink Methods

Methods for calculating the directional hyperlink constitute the second category in the present categorization. A directional hyperlink can be generally applied to all social interaction scenarios, where different participants have different social roles. The simplest method is probably the cross-correlation analysis, which measured the time lagged Pearson's correlation between two datasets (e.g. [29, 35]. The most popular methods, however, are those based on *multivariate auto-regression model* (MVAR), e.g. Granger's causality (GC), directed transfer function (DTF), partial directed coherence (PDC), etc. The MVAR model describes the underlying order of a multivariate data by modelling the current value of the variables as a weighted linear sum of all the previous values. In the present hyperlink context, it can be formulated as below

$$\begin{bmatrix} x_{Ch1}^{S1}(n) \\ x_{Ch2}^{S1}(n) \\ \vdots \\ x_{ChM}^{SN}(n) \end{bmatrix} = \sum_{i=1}^{m} A_i \begin{bmatrix} x_{Ch1}^{S1}(n-i) \\ x_{Ch2}^{S1}(n-i) \\ \vdots \\ x_{ChM}^{SN}(n-i) \end{bmatrix} + \begin{bmatrix} w_{Ch1}^{S1}(n) \\ w_{Ch2}^{S1}(n) \\ \vdots \\ w_{ChM}^{SN}(n) \end{bmatrix}, \tag{10.8}$$

where $x_{Chj}^{Sk}(n)$ is the EEG data from participant *Sk*, at channel *j* and sampling point *n*; A_i is coefficient matrix for time lag *i*; and *w* is the prediction error. MVAR characterizes dependencies within the multivariate data (i.e. the multi-channel EEG data from multiple participants), specifically in terms of the historical influence of one variable on another. The MVAR-based methods hereby yield results in the form of neural connectivity patterns across different brain regions, both within and across participants. One necessary preprocessing procedure for hyperlink-based MVAR is a within-participant normalization before pooling all data together, considering the inter-participant differences in EEG signals [63]. Among these MVAR-based methods, PDC has been suggested to be of particular interest, as it can distinguish between direct and indirect connectivity flows in the estimated connectivity pattern [47]. Using such methods, significant directional hyperlinks have been reported in a number of studies (e.g. [1, 53, 63]). Many other advanced EEG signal processing methods, such as mutual information, entropy and so on, however, have not been widely applied for analyzing EEG-based hyperlinks. Considering their successes in single-brain analysis (see [40, 59] for reviews), these methods are expected to help us better model hyperlinks and therefore further extend our understanding of social neuroscience.

10.3.3 Machine-Learning Methods for Hyperlinks

In contrast to the above two categories, machine-learning methods aim at predicting certain behavioral or mental states on the basis of multi-brain data. The rationale behind this approach is: Different social conditions are expected to be linked with distinguishable hyperlink patterns; machine-learning methods therefore can computationally learn the condition-specific patterns. Machine-learning methods can be applied on both the raw multi-brain data, or extracted neural features, e.g., by using the above-mentioned two types of hyperlink methods as well. Popular machine-learning methods for neural signal processing include linear discriminant analysis, support vector machine, random forest, etc. [36]. These methods have been widely used for decoding different mental states for individuals, towards brain-computer interface applications [16, 67] as well as basic neuroscience research [25]. Compared to the unidirectional and directional hyperlink methods, machine-learning methods output predictive models that do not necessarily fit the underlying neurophysiological properties of our brain. Even if these models can effectively predict different behavioral or mental states, they are not informative about the underlying neural mechanism by themselves, i.e. lacking explanatory power [55]. Nevertheless, machine-learning methods are advantageous when mining vast amounts of high-dimensional data, as in the case of multi-brain analysis [33, 46].

Researchers are starting to take the machine-learning approach for analyzing multi-brain EEG data. Movement directions can be predicted with accuracies from 66% in a single-brain condition to 95% when data from 20 brains were included [66]. Enhanced perceptual decision accuracy was achieved by aggregating EEG activities from a group of participants, for both a discrimination task [15] and a visual search task [64]. In a more interactive scenario, frontal alpha oscillations could effectively distinguish between leaders from followers involved in a motor coordination task [34]. More studies in this direction are expected to emerge in the near future, translating the already reported hyperlink findings into predictive power and thus facilitating hyperlink-based applications, e.g., evaluating and training of social interaction abilities.

10.4 Future Perspectives

The development of hyperscanning methods is at its early beginning. While the hyperscanning technique is bringing exciting findings to the field of social neuroscience, critical methodological challenges remain to be further addressed, as summarized below.

First, EEG recordings in naturalistic social interaction environment will inevitably be affected by extensive artifacts due to necessary movements of the eyes, faces, limbs etc., as well as electrical noises in the normally unshielded environment. Therefore, artifact rejection need to be treated with high priority and caution need to be taken when validating its effectiveness. Whereas conventional artifact rejection methods targeting at modelling environmental or physiological noises can be readily applied, more advanced methods are needed to be developed, in order to better remove the possibly stronger artifacts during social interactions. For instance, it has recently been demonstrated that artifact reduction could be achieved by modelling the valid signals [5, 6, 11, 42].

Second, a new methodological framework need to be defined in order to deal with data coming from different brains. Most of the EEG signal processing methods to date are based on the assumption that the neural signals are generated by the same system (i.e. brain). Although progress has been to address this issue by normalization of the individual brain's data, or extraction of non-individual-specific information [2], these latest methods work mainly within the signal space, on the basis of temporal, spectral or spatial features. In other words, the reported hyperlinks imply similarities across individuals in the EEG signal space. Possible similarity in higher cognitive levels, has not been systematically investigated. Nevertheless, the success of representational similarity analysis for visual objection recognition in fMRI provide strong evidence for the existence of such across-individual similarity

[24]: different brains might encode stimuli in different ways, but the mental distances among these encoded stimuli are expected to remain largely invariant. Therefore, it is reasonable to hypothesize the existence of inter-brain hyperlinks in a higher-level representational space. As such a representational space is more closely related to our mental world than the signal space, representational hyperlinks should have more important theoretical implications, as compared to our current findings.

Last but not least, the latest development in artificial intelligence and machine learning methods, such as the deep learning neural networks, may help facilitate our exploration of multi-brain data [20, 50]. These methods are expected to extract critical information from high-dimensional multi-brain data without explicit modelling and extensive labor, speeding up the development of the social neuroscience field.

References

1. F. Babiloni, F. Cincotti, D. Mattia et al., Hypermethods for EEG hyperscanning, in *28th Annual International Conference of the IEEE Engineering in Medicine and Biology Society* (New York, NY, USA, 30 August–3 September 2006)
2. F. Babiloni, L. Astolfi, Social neuroscience and hyperscanning techniques: past, present and future. Neurosci. Biobehav. Rev. **44**, 76–93 (2014)
3. D.A. Bridwell, C. Roth, C.N. Gupta, V.D. Calhoun, Cortical response similarities predict which audiovisual clips individuals viewed, but are unrelated to clip preference. PLoS ONE **10**(6), e0128833 (2015)
4. W.-T. Chang, I.P. Jääskeläinen, J.W. Belliveau et al., Combined MEG and EEG show reliable patterns of electromagnetic brain activity during natural viewing. NeuroImage **114**, 49–56 (2015)
5. A. de Cheveigné, L.C. Parra, Joint decorrelation, a versatile tool for multichannel data analysis. NeuroImage **98**, 487–505 (2014)
6. A. de Cheveigné, Sparse time artifact removal. J. Neurosci. Methods **262**, 14–20 (2016)
7. H. De Jaegher, E. Di Paolo, R. Adolphs, What does the interactive brain hypothesis mean for social neuroscience? A dialogue. Phil. Trans. R. Soc. B **371**(1693), 20150379 (2016)
8. E.A. Di Paolo, H. De Jaegher, The interactive brain hypothesis. Front. Hum. Neurosci. **6**, 163 (2012)
9. J.P. Dmochowski, P. Sajda, J. Dias, L.C. Parra, Correlated components of ongoing EEG point to emotionally laden attention–a possible marker of engagement? Front. Hum. Neurosci. **6**, 112 (2012)
10. J.P. Dmochowski, M.A. Bezdek, B.P. Abelson et al., Audience preferences are predicted by temporal reliability of neural processing. Nat. Commun. **5**, 4567 (2014)
11. J.P. Dmochowski, A.S. Greaves, A.M. Norcia, Maximally reliable spatial filtering of steady state visual evoked potentials. NeuroImage **109**, 63–72 (2015)
12. T.D. Duane, T. Behrendt, Extrasensory electroencephalographic induction between identical twins. Science **150**(3694), 367 (1965)
13. G. Dumas, J. Nadel, R. Soussignan et al., Inter-brain synchronization during social interaction. PLoS ONE **5**(8), e12166 (2010)
14. G. Dumas, F. Lachat, J. Martinerie et al., From social behaviour to brain synchronization: review and perspectives in hyperscanning. Irbm **32**(1), 48–53 (2011)
15. M.P. Eckstein, K. Das, B.T. Pham et al., Neural decoding of collective wisdom with multi-brain computing. NeuroImage **59**(1), 94–108 (2012)
16. S. Gao, Y. Wang, X. Gao, B. Hong, Visual and auditory brain–computer interfaces. IEEE Trans. Biomed. Eng. **61**(5), 1436–1447 (2014)

17. G.H. Golub, C.F. Van Loan, *Matrix Computations* (Johns Hopkins University Press, Baltimore, 1996)
18. N. Hachmeister, A. Finke, H. Ritter, Does machine-mediated interaction induce inter-brain synchrony?—A hyperscanning study, in *36th Annual International Conference of the IEEE Engineering in Medicine and Biology Society, Chicago, IL, USA*, 26–30 August 2014
19. R. Hari, L. Henriksson, S. Malinen, L. Parkkonen, Centrality of social interaction in human brain function. Neuron **88**(1), 181–193 (2015)
20. D. Hassabis, D. Kumaran, C. Summerfield, M. Botvinick, Neuroscience-inspired artificial intelligence. Neuron **95**(2), 245–258 (2017)
21. U. Hasson, Y. Nir, I. Levy et al., Intersubject synchronization of cortical activity during natural vision. Science **303**(5664), 1634–1640 (2004)
22. U. Hasson, R. Malach, D.J. Heeger, Reliability of cortical activity during natural stimulation. Trends. Cogn. Sci. **14**(1), 40–48 (2010)
23. U. Hasson, A.A. Ghazanfar, B. Galantucci et al., Brain-to-brain coupling: a mechanism for creating and sharing a social world. Trends. Cogn. Sci. **16**(2), 114–121 (2012)
24. J.V. Haxby, A.C. Connolly, J.S. Guntupalli, Decoding neural representational spaces using multivariate pattern analysis. Annu. Rev. Neurosci. **37**(1), 435–456 (2014)
25. J.-D. Haynes, G. Rees, Decoding mental states from brain activity in humans. Nat. Rev. Neurosci. **7**, 523–534 (2006)
26. J. Jahng, J.D. Kralik, D.-U. Hwang, J. Jeong, Neural dynamics of two players when using nonverbal cues to gauge intentions to cooperate during the Prisoner's Dilemma Game. NeuroImage **157**, 263–274 (2017)
27. D.-Y. Kang, J. Kim, Y.-S. Shin et al., Coherent neural responses of human populations during watching movie, in *14th International Conference on Control, Automation and Systems, Seoul, South Korea*, 22–25 October 2014
28. A.Y. Kaplan, A.A. Fingelkurts, A.A. Fingelkurts et al., Nonstationary nature of the brain activity as revealed by EEG/MEG: methodological, practical and conceptual challenges. Sig. Process. **85**(11), 2190–2212 (2005)
29. M. Kawasaki, Y. Yamada, Y. Ushiku et al., Inter-brain synchronization during coordination of speech rhythm in human-to-human social interaction. Sci. Rep. **3**, 1692 (2013)
30. J.J. Ki, S.P. Kelly, L.C. Parra, Attention strongly modulates reliability of neural responses to naturalistic narrative stimuli. J. Neurosci. **36**(10), 3092–3101 (2016)
31. B. King-Casas, D. Tomlin, C. Anen et al., Getting to know you: reputation and trust in a two-person economic exchange. Science **308**(5718), 78–83 (2005)
32. T. Koike, H.C. Tanabe, N. Sadato, Hyperscanning neuroimaging technique to reveal the "two-in-one" system in social interactions. Neurosci. Res. **90**, 25–32 (2015)
33. I. Konvalinka, A. Roepstorff, The two-brain approach: how can mutually interacting brains teach us something about social interaction? Front. Hum. Neurosci. **6**, 215 (2012)
34. I. Konvalinka, M. Bauer, C. Stahlhut et al., Frontal alpha oscillations distinguish leaders from followers: multivariate decoding of mutually interacting brains. NeuroImage **94**, 79–88 (2014)
35. A.K. Kuhlen, C. Allefeld, J.-D. Haynes, Content-specific coordination of listeners' to speakers' EEG during communication. Front. Hum. Neurosci. **6**, 266 (2012)
36. S. Lemm, B. Blankertz, T. Dickhaus, K.-R. Müller, Introduction to machine learning for brain imaging. NeuroImage **56**(2), 387–399 (2011)
37. M.D. Lieberman, Social cognitive neuroscience: a review of core processes. Annu. Rev. Psychol. **58**, 259–289 (2007)
38. T. Liu, M. Pelowski, A new research trend in social neuroscience: Towards an interactive-brain neuroscience. PsyCh J. **3**(3), 177–188 (2014)
39. P.R. Montague, G.S. Berns, J.D. Cohen et al., Hyperscanning: simultaneous fMRI during linked social interactions. NeuroImage **16**(4), 1159–1164 (2002)
40. S. Motamedi-Fakhr, M. Moshrefi-Torbati, M. Hill et al., Signal processing techniques applied to human sleep EEG signals—a review. Biomed. Signal Process. Control **10**, 21–33 (2014)
41. V. Müller, J. Sänger, U. Lindenberger, Intra-and inter-brain synchronization during musical improvisation on the guitar. PLoS ONE **8**(9), e73852 (2013)

42. M. Nakanishi, Y. Wang, X. Chen et al., Enhancing detection of SSVEPs for a high-speed brain speller using task-related component analysis. IEEE Trans. Biomed. Eng. **65**(1), 104–112 (2018)
43. J. Pajula, J. Tohka, How many is enough? Effect of sample size in inter-subject correlation analysis of fMRI. Comput. Intell. Neurosci. **2016**, 2094601 (2016)
44. L. Parra, P. Sajda, Blind source separation via generalized eigenvalue decomposition. J. Mach. Learn. Res. **4**(Dec), 1261–1269 (2003)
45. A. Pérez, M. Carreiras, J.A. Duñabeitia, Brain-to-brain entrainment: EEG interbrain synchronization while speaking and listening. Sci. Rep. **7**(1), 4190 (2017)
46. R. Raina, Y. Shen, A. Mccallum, A.Y. Ng, Classification with hybrid generative/discriminative models, in *Advances in neural information processing systems 16, Vancouver and Whistler, British Columbia, Canada*, 8–13 December 2003 (2004)
47. K. Sameshima, L.A. Baccalá, Using partial directed coherence to describe neuronal ensemble interactions. J. Neurosci. Methods **94**(1), 93–103 (1999)
48. P. Sauseng, W. Klimesch, What does phase information of oscillatory brain activity tell us about cognitive processes? Neurosci. Biobehav. Rev. **32**(5), 1001–1013 (2008)
49. L. Schilbach, B. Timmermans, V. Reddy et al., Toward a second-person neuroscience 1. Behav. Brain Sci. **36**(4), 393–414 (2013)
50. R.T. Schirrmeister, J.T. Springenberg, L.D.J. Fiederer et al., Deep learning with convolutional neural networks for EEG decoding and visualization. Hum. Brain Mapp. **38**(11), 5391–5420 (2017)
51. R. Schmälzle, F.E. Häcker, C.J. Honey, U. Hasson, Engaged listeners: shared neural processing of powerful political speeches. Soc. Cogn. Affect. Neurosci. **10**(8), 1137–1143 (2015)
52. C.E. Schroeder, P. Lakatos, Y. Kajikawa et al., Neuronal oscillations and visual amplification of speech. Trends. Cogn. Sci. **12**(3), 106–113 (2008)
53. N. Sciaraffa, G. Borghini, P. Aricò et al., Brain interaction during cooperation: evaluating local properties of multiple-brain network. Brain Sci. **7**(7), 90 (2017)
54. X. Shen, Q. Sun, Y. Yuan, A unified multiset canonical correlation analysis framework based on graph embedding for multiple feature extraction. Neurocomputing **148**, 397–408 (2015)
55. G. Shmueli, To explain or to predict? Stat. Sci. **25**(3), 289–310 (2010)
56. M. Siegel, T.H. Donner, A.K. Engel, Spectral fingerprints of large-scale neuronal interactions. Nat. Rev. Neurosci. **13**(2), 121–134 (2012)
57. L.J. Silbert, C.J. Honey, E. Simony et al., Coupled neural systems underlie the production and comprehension of naturalistic narrative speech. Proc. Natl. Acad. Sci. **111**(43), E4687–E4696 (2014)
58. N. Sinha, T. Maszczyk, Z. Wanxuan et al., EEG hyperscanning study of inter-brain synchrony during cooperative and competitive interaction, in *2016 IEEE International Conference on Systems, Man, and Cybernetics* (2016)
59. C.J. Stam, Nonlinear dynamical analysis of EEG and MEG: review of an emerging field. Clin. Neurophysiol. **116**(10), 2266–2301 (2005)
60. D.A. Stanley, R. Adolphs, Toward a neural basis for social behavior. Neuron **80**(3), 816–826 (2013)
61. G.J. Stephens, L.J. Silbert, U. Hasson, Speaker–listener neural coupling underlies successful communication. Proc. Natl. Acad. Sci. **107**(32), 14425–14430 (2010)
62. C. Szymanski, A. Pesquita, A.A. Brennan et al., Teams on the same wavelength perform better: Inter-brain phase synchronization constitutes a neural substrate for social facilitation. NeuroImage **152**, 425–436 (2017)
63. J. Toppi, G. Borghini, M. Petti et al., Investigating cooperative behavior in ecological settings: an EEG hyperscanning study. PLoS One **11**(4), e0154236 (2016)
64. D. Valeriani, R. Poli, C. Cinel, Enhancement of group perception via a collaborative brain—computer interface. IEEE Trans. Biomed. Eng. **64**(6), 1238–1248 (2017)
65. F. van Overwalle, Social cognition and the brain: a meta-analysis. Hum. Brain Mapp. **30**(3), 829–858 (2009)

66. Y. Wang, T.-P. Jung, A collaborative brain-computer interface for improving human performance. PLoS ONE **6**(5), e20422 (2011)
67. J. Wolpaw, E.W. Wolpaw, *Brain-Computer Interfaces: Principles and Practice* (Oxford University Press, New York, 2012)
68. D. Zhang, B. Hong, S. Gao, B. Röder, Exploring the temporal dynamics of sustained and transient spatial attention using steady-state visual evoked potentials. Exp. Brain Res. **235**(5), 1575–1591 (2017)
69. J.R. Zhang, J. Sherwin, J. Dmochowski et al., Correlating speaker gestures in political debates with audience engagement measured via EEG. In: Proceedings of the 22nd ACM international conference on multimedia, Orlando, Florida, USA, 3–7 November 2014 (2014)
70. Y. Zhang, G. Zhou, J. Jin et al., Sparse Bayesian multiway canonical correlation analysis for EEG pattern recognition. Neurocomputing **225**, 103–110 (2017)

The manufacturer's authorised representative in the EU is Springer Nature Customer Service Centre GmbH, Europaplatz 3, 69115 Heidelberg, Germany. If you have any concerns regarding our products, please contact ProductSafety@springernature.com

Printed and bound by CPI Group (UK) Ltd, Croydon, CR0 4YY
25/09/2026
02222892-0007